AMERICAN LOCOMOTIVES
OF THE MIDLAND
RAILWAY

AMERICAN LOCOMOTIVES OF THE MIDLAND RAILWAY

by
DAVID HUNT

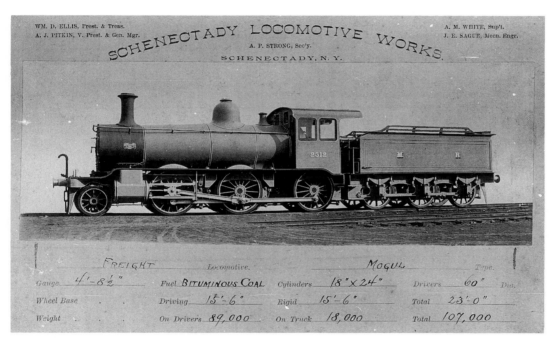

FREIGHT Locomotive.		MOGUL Type.	
Gauge 4'-8½"	Fuel BITUMINOUS COAL	Cylinders 18"x24"	Drivers 60" Dia.
Wheel Base	Driving 15'-6"	Rigid 15'-6"	Total 23'-0"
Weight	On Drivers 89,000	On Truck 18,000	Total 107,000

A Schenectady Works description with a photograph of the second Midland Railway mogul built by that firm. Some of the leading details are given below the picture, including the fact that the driving wheels were 5ft 0in diameter and not, as some previous writers have claimed, 5ft 1in. COLLECTION R. J. ESSERY

WILD SWAN PUBLICATIONS

CONTENTS

© Wild Swan Publications Ltd. and David Hunt 1997
ISBN 1 874103 41 0

Designed by Paul Karau
Printed by Amadeus Press Ltd., Huddersfield

Published by
WILD SWAN PUBLICATIONS LTD.
1-3 Hagbourne Road, Didcot, Oxon, OX11 8DP

INTRODUCTION

The Midland Railway was the major purchaser of American-built locomotives in the United Kingdom. In making this statement I am including engines bought by the Birmingham & Gloucester Railway before it became a part of the Midland in 1846. To be precise, the Birmingham & Gloucester amalgamated with the Bristol & Gloucester Railway in 1845 to form the Bristol & Birmingham Railway, which was operated by a joint management committee; it was decided, however, that the constituent companies would amalgamate with the Midland Railway as separate entities, as would the joint undertaking, and so it was that, in theory, three concerns were absorbed into the Midland Railway on 3rd of August 1846. Counting these Birmingham & Gloucester/Bristol & Birmingham locomotives, the history of Midland Railway motive power includes a total of fifty-seven American-built engines. Some writers in the past have put the total higher than this, but, as I hope to show later, they have included in their reckoning some close copies of the Birmingham & Gloucester's Norris engines built by English firms. Since these latter locomotives were, to all intents and purposes, American designs, however, I have included them in my coverage of the true Transatlantic machines in this publication. Their inclusion raises the total of American-built or inspired locomotives owned by the Midland Railway or its constituent companies to sixty-six.

Few as their numbers were, however, these engines have been the subject of much controversy over the years as well as a lot of myth, confusion, and downright inaccuracy. In this work I hope to produce an accurate account of the subject and present a consolidated history of American locomotives on the Midland Railway.

Some of the myths and legends surrounding these machines have been started or perpetuated by well-known authorities on the steam locomotive — even Ahrons got some things wrong when writing about them[1] — and unravelling fact, fiction and guesswork is not easy. In 1908 and 1909 R.M. Deeley, then locomotive superintendent of the Midland Railway, complicated the issue by providing lists, purported to be of Birmingham & Gloucester Railway locomotives, to *Locomotive Magazine* and *Railway Magazine*, but they were inaccurate as he was only referring to engines taken into Midland Railway stock; even then there were errors in his reckoning. This led to years of confusion.

I first became interested in the later American locomotives from Baldwins and the Schenectady works over twenty years ago and have been collecting information on them ever since; later I included the earlier Norris engines in my researches. Two of the most fruitful sources of material on the various types were the late Laurie Ward and the papers of P.C. Dewhurst. Originally it was the intention of both myself and the Editor that the story of these engines should form part of Volume 5 of *An Illustrated Review of Midland Locomotives* to be published by Wild Swan; they did not, however, seem to fit in to the Kirtley story and when, due to unforeseen circumstances which will, I hope, be resolved soon, Volume 5 was delayed, we decided that *Midland Record* would be a good vehicle.

The story of American locomotives on the Midland Railway covers the first two periods of Midland Railway history as delineated by the late Bill Steele and described in the Preview Edition of *Midland Record*, namely period one prior to 1844 and period two up to 1923. Because of a gap of nearly fifty years between the disappearance of the earlier Norris engines from Midland metals and the acquisition of the Baldwin and Schenectady moguls, I have divided the work into two chapters.

As with other articles I have written, I have done my best to get it right, but infallibility and omniscience are not among my attributes. As the Editor of *Midland Record* has said to me many times, though, "If you wait until every 'i' is dotted and every 't' crossed, you'll probably die before you've had anything published." So, here is my attempt at the story of Midland Railway American locomotives, and if you can add or correct anything please let me know or write to the Editor.

---

1. E.L. Ahrons, *The British Steam Locomotive From 1825 to 1925*, published in 1927. In saying that Ahrons was wrong about some aspects of these engines, I mean that he made statements which are not borne out by any other source I have found. Other authors writing about Midland locomotives, such as C.E. Stretton and Hamilton Ellis, have also made errors when writing about the Norris engines.

A painting by F. J. Dolby showing Norris 'A Extra' 4–2–0 Philadelphia ascending the Lickey incline with a train of eight loaded mineral wagons in 1841 or 1842. Although obviously inaccurate dimensionally, the painting shows the sanding apparatus fitted to the engine in late 1840, and a forward safety valve on the boiler barrel. The dumb buffers, raised above the level of the front beam, can also be seen. Note that at this time the track on the Lickey was laid on a Brunel-style baulk road.
COLLECTION R. J. ESSERY

CHAPTER ONE
THE NORRIS LOCOMOTIVES

WHEN Isambard Kingdom Brunel surveyed a line of railway from Birmingham to Gloucester in 1832, the directors of the company ordering the survey dictated that it should be built as cheaply as possible and that it should avoid large towns en route — a curious principle, one may think, for a commercial railway undertaking, but one driven by the cost of land and construction. The history of the line before the Midland Railway took over was always dogged by lack of money, and for a long time after Brunel carried out his survey nothing was done apart from trying to secure financial backing for the undertaking. Brunel's route from Gloucester to Birmingham would have taken the line to the east of that eventually adopted and secured a ruling gradient of one in three hundred; generations of enginemen, operating staff, and locomotive superintendents must have wished that his proposals had been adopted. The problem, though, was that the suggested route was so far from population centres that the promoters were forced to alter it in order to gain enough financial support to go ahead with their project. By the time they had done this, Brunel had gone to the Great Western and a new engineer had to be found to take the scheme forward.

Locke, in engineering the Lancaster & Carlisle Railway, had ignored the pronouncements of the Stephensons about gradients and had taken his line over Shap, confident in the development of the locomotive being able to overcome the obstacle. Prophetically, he also stated that, in his view, braking would prove to be a greater problem than tractive effort. The Birmingham & Gloucester's engineer, Captain W.S. Moorsom, went one better and, in looking for a suitable route, he decided to go through the Lickey hills and built his line containing one of the greatest operating headaches ever to bedevil a main line British railway — the two and a bit miles of 1 in 37 Lickey incline from Bromsgrove to Blackwell. Throughout the days of steam traction it continued to provide a spectacle for the enthusiast and a problem for railwaymen both going up and coming down, as the Editor can vouch from his days on the footplate.

The Midland Railway is reputed to have been considering electric traction for the line prior to the First World War in the hope of alleviating their problems, but the constraints on development caused by that most insane of events put paid to such schemes. When it was being built, the Lickey incline was declared by both Brunel and George Stephenson to be unworkable by locomotive haulage, a pronouncement which must have caused the Birmingham & Gloucester's directors some reflection on their engineer's choice of route. Nevertheless, Moorsom went ahead with his plan and the line was driven straight up that fearsome slope, and therein lay the beginnings of the American engine story.

THE *GEORGE WASHINGTON* AND THE SHUYLKILL TRIALS

Moorsom's approach to the question of motive power for his line seems to have been one of hopeful pragmatism, and when he was told by the directors in 1838 to have locomotives ready for the opening of the line, he ordered some from British manufacturers which could never hope to pull anything worthwhile up the Lickey. During his deliberations,

however, he became aware of stories emanating from across the Atlantic concerning prodigious feats of hill-climbing by a locomotive built at the Bush Hill works of Norris in Philadelphia; in this he probably saw his salvation. I have seen it written[2] that Moorsom actually visited America and saw Norris engines in action, but have been unable to confirm this through reliable evidence.

The subject of these tales was the *George Washington*, a 4-2-0 produced by Norris in July 1836 and sent to work on the Philadelphia & Columbia Railroad. This line included a truly awesome inclined plane at Shuylkill, or Belmont as it is sometimes referred to, just outside Philadelphia, which was cable hauled — no wonder as its half mile length was at an inclination of just under one in fourteen! It was up this hill that the *George Washington* hauled 8 tons 11 cwt (including tender) in two minutes and one second on 10th July 1836, and nine days later dragged the then amazing load of just under fourteen tons to the top in only twenty-three seconds more. But Norris had cheated, or at least misled observers, and Moorsom was deceived (he wasn't alone). The engine weighed 14,930 pounds, of which 8,700 pounds was on the driving wheels, and the boiler pressure was given as 'just under sixty pounds' on 10th July, and 'just under eighty pounds' on the 19th. With the combination of this adhesive weight and the tractive effort produced by the reported boiler pressure, the laws of physics make it impossible for such feats of haulage to have been achieved. During the first trial a boiler pressure of only sixty pounds per square inch simply would not have been enough to provide the tractive effort required, and, even if the required power had been available, the engine would have been on the point of slipping as it ascended the incline. The later trial would have been even worse as, even with the power output increased to the point where the locomotive could achieve the necessary effort, the low adhesive weight would have prevented it from climbing the slope; it would have slipped long before the required drawbar pull had been developed. As a result of the reports he received of the trials, however, Moorsom entered into negotiations on 3rd September 1838 with Norris for the purchase of locomotives, and in due course reported to the directors that in a letter dated 15th October 1838, 'Mr Norris offers to supply an assistant engine for working up the inclined plane', and that, subject to satisfactory trials in England, ten more of the same class would be contracted for. The directors seemed happy with this arrangement and requested Moorsom to attend a meeting in company with Mr. Gwynne, Norris's agent in England, to ratify the situation.

So how, given that it would seem to be impossible, did the *George Washington* achieve its trial results on the Shuylkill incline? For there is no doubt from the contemporary

2. The earliest reference I have seen to Moorsom's supposed visit to America is in *The Midland Railway — Its Rise and Progress* by F.S. Williams, published in 1875, in which he writes, 'Captain Moorsom, however, when in America, had seen . . .', but gives no further details. There is no reference in Birmingham and Gloucester records of such a visit and no other writer gives any more information. Williams also wrote that twelve or fourteen engines were initially ordered, that the driving wheels were only two feet in diameter, and that, 'On arrival in this country, they did all that was expected of them'.

NORRIS 4–2–0 *GEORGE WASHINGTON*

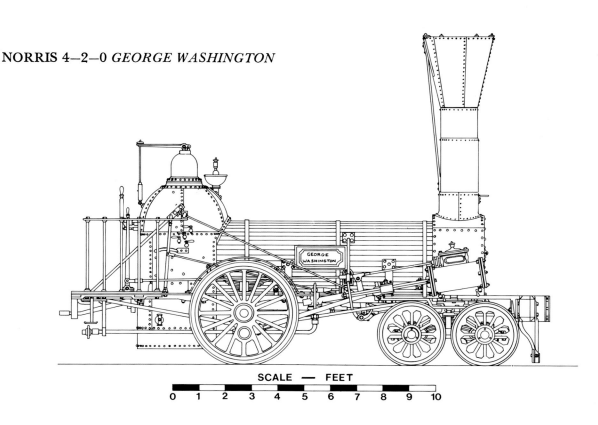

SCALE — FEET

0 1 2 3 4 5 6 7 8 9 10

Side elevation of George Washington, the 1836 Norris 4–2–0 which drew Captain Moorsom's attention following its hill-climbing feats at Schuylkill. The engine was not of what later became regarded and written about as the 'true' Norris type as it had the connecting rods positioned inside the driving wheels acting on half cranks between the wheels and the axleboxes. The centres of the leading wheels were wood fitted with chilled cast-iron tyres. The dropped drawbar, which helped to increase the locomotive's adhesive weight, can be seen angled down from the platform. The boiler was lagged with pine slats but the firebox was unlagged. In most respects, however, George Washington was very similar to later Norris standard types of 4–2–0. The original drawing from which the diagram was developed was by J. Snowdon Bell.

reports that it did.[3] Some years ago I was fortunate enough to obtain a copy of a paper presented by the late P.C. Dewhurst to the Institution of Civil Engineers on 15th October 1947, in which he analysed the available evidence quite brilliantly. As regards the first test, the locomotive could have only produced the require tractive effort at over seventy-six pounds per square inch boiler pressure and, even if this had been the case, the adhesive weight would have been barely sufficient. The second test would have needed over one hundred pounds per square inch boiler pressure, but the adhesive weight required was 12,638 pounds, clearly well above the figures quoted for the locomotive. Taking these points in reverse order, and accepting the loads and engine weight as being accurate due to the number of independent engineers and other witnesses present, Dewhurst first addressed the adhesion weight problem as having the greatest bearing on the second test. No matter what the power output, the *George Washington* could not have pulled fourteen tons up a one in fourteen incline with only slightly under four tons resting on the driving wheels. He realised, however, that the true figure was much higher than this. In designing the locomotive, Norris and his chief designer, Fred

C. Sanno, had arranged the drawbar at such a low level that a proportion of the weight on the front of the tender would be transferred to the driving wheels, as would any extra weight carried on the overhanging rear platform of the engine. From evidence of the trials, it became apparent that the tender and footplate of the *George Washington* had been somewhat crowded, and Dewhurst was able to show that three men on the tender and another three on the locomotive would transfer sufficient weight to the driving wheels to make the adhesion increase by no less than 1,776 pounds. Added to this, the effect of the drawbar design would have transferred some of the locomotive's tractive effort into a downward thrust on the rear of the engine, producing a further 923 pounds of adhesive weight. The shortfall still required he explained by a combination of water in the boiler being displaced to the rear because of the incline, thus throwing even more weight onto the driving wheels, and the tendency, when steaming hard, to lift the front end of the

3. The trial of 10th July 1836 was reported fully in the *American Railroad Journal* on 16th July, and the later trial of 19th July covered in the same magazine on the 30th.

locomotive and so bear down on the drivers. These calculations are borne out by a Norris employee writing in 1876 that:

'Extra weight was thrown on the drivers by the action of the draft link connecting the tender with the engine, the result being that about all the weight of the locomotive rested on the driving wheels.'

Further evidence was given by Moorsom, stating in 1840 that:[4]

'The mode of attaching the tender was peculiar . . . it threw a portion of the weight of the tender on the engine . . . an advantage in starting.'

So the adhesion weight problem is explicable, but what of the boiler pressure needed to provide the requisite tractive effort? To solve this problem, Dewhurst turned to the safety valve, which on the *George Washington* was a single spring balance type. A study of contemporary views showed him that the proportions of the lever and likely area of the valve would result in a boiler pressure three times that of the reaction on the end of the lever. Even if this was corrected by calibrating the reading at twenty pounds reaction, ie, sixty pounds per square inch true boiler pressure, the boiler pressures at indications of seventy and eighty pounds per square inch would actually be ninety and one hundred and twenty pounds per square inch respectively. As the graduations on the lever appeared from illustrations of the locomotive to be linear, it is certain that errors of this sort would have occurred in the apparent boiler pressures on the gauge. In addition, the design of the valve would mean that the actual blowing-off pressure was about five pounds per square inch above that shown on the gauge, even if it was accurate. This view is supported by contemporary doubts expressed in England about the *George Washington*'s true boiler pressure, and by McConnell saying to the Institution of Mechanical Engineers in 1849:

'The pressure, however, on the American engines was very fallacious, for the spring balance only indicated about one-third of the actual pressure in the boiler, which was really about one hundred pounds per inch.'

Also, in 1840, the Birmingham & Gloucester stipulated to Nasmyths when ordering some Norris type locomotives that:

'The engines shall be made like the sample engine . . . and that the scale of each of the (two) valves shall be graduated to show directly in figures the pressure of steam in the boiler per square inch and not in triplicate proportion.'

NEGOTIATIONS AND TRIALS IN BRITAIN

That the engines were not capable of anything like their accredited performance if the weights and steam pressure had been as advertised was, to Moorsom's chagrin, borne out by trials in England. The *George Washington* had been superseded three months later by the first of the 'true' Norris engines, the *Washington County Farmer*, which also operated on the Philadelphia & Columbia and succeeded in eclipsing its predecessor's performance.[5] The *Washington County Farmer*'s feats seem to have been achieved by dint of the same sleight of hand. To be charitable, it is possible of course that the 'fiddles' which enabled the Norris engines to achieve their performances were unintentional. This engine, as Dewhurst has shown,[6] was the first 'true' Norris type as eventually purchased by the Birmingham & Gloucester Railway. The salient features of the type were as follows: outside cylinders and connecting rods, leading four wheeled bogie, bar frames, and single driving axle positioned forward of the firebox.

The meeting between the directors and Norris's agent, Gwynne, took place on 15th November 1838 and the minutes record the following:

'The agent of Mr. Norris of Philadelphia, America, having proposed to furnish *two* Engines to the Birmingham and Gloucester Railway Company to perform as follows:

'For the Inclined Plane two Miles and sixty six Yards. To ascend Inclined Plane of one in thirty eight.

'To carry loads of seventy five tons including Engine and Tender up such a plane at speed of fifteen miles per hour. Weight of Engine about twelve tons.

'For the General Line, inclination one in three hundred. To carry load of one hundred tons, including Engine and Tender, upon such a line at speed of twenty miles per hour. Weight of Engine about eight tons.

'Subject to trial in this Country as to both Engines on the terms stated in Mr. Norris's letter and if approved two more of the large engines to be had and six smaller at the same price and subject to the same trial and terms of payment. Resolved that this offer be accepted.'

This is the first reference I have seen to two engines being sent for trial, and can only assume that Norris or Gwynne persuaded the directors (or Gloucester Committee as they were then known) to take them. It seems puzzling, though, as the smaller one was a 'B' type which had not featured either in Moorsom's initial approach or in Norris's original offer, and the larger engine would have been far more appropriate even for the less demanding parts of the line. The stipulation that the 'Assistant engine' be capable of ascending the Lickey incline at 15mph hauling 75 tons was acknowledged in Norris's letter to Moorsom on 15th October but no mention had been made at that stage of anything other than a 12 ton locomotive nor of any other performance target.

A story has grown up that the reason Moorsom was forced to turn to America for engines to work the Lickey was that English firms, such as Stephensons, could not build locomotives of sufficient power. This seems to be a result of Moorsom's statement in 1839 that 'It was not until two celebrated makers in England had refused to undertake an order that negotiations were set in foot with Mr Norris.'

---

4. Moorsom presented a paper to the Institution of Civil Engineers in 1840 entitled 'Experiments on Locomotive Engines Manufactured by Mr. Norris' in which he gave results of the trials conducted in this country.

5. The *Pennsylvania Enquirer* of 20th October 1836 reported that two days earlier the *Washington County Farmer* had pulled a load of 24½ tons up the Shuylkill incline in three minutes and fifteen seconds. In the *Washington County Farmer* the inside half cranks of the *George Washington* were replaced by outside rods driving directly onto the driving wheel bosses. For many years it was assumed by railway historians that the *George Washington* was the first of the 'true' Norris engines, but the American J. Snowdon Bell, and later P.C. Dewhurst, showed otherwise.

6. P.C. Dewhurst, *The Norris Locomotive in the U.S.A.*, Railway and Locomotive Historical Society 1945.

NORRIS 'B' TYPE

An estimation of one of the Birmingham & Gloucester Railway's Norris 'B' type 4–2–0s. Like George Washington, these engines had pine slats around the boiler but the fireboxes were unlagged and were of the true Bury type with a 'D'-shaped hearth and the top drawn up into a hemisphere; not all Norris engines had the true Bury firebox, some were of circular section at the hearth. This diagram is an attempt to show what the Birmingham & Gloucester Norris 'B' type locomotives may have looked like, and as no drawing of them has survived, I have based it on two main sources:

A plate from a Norris brochure of 1838 showing a standard 'B' type contemporary with Moorsom's approach to Norris.

A drawing of a 'B' type supplied to the Berlin–Potsdam Railway in 1839, which appeared in Die Entwicklung Der Lokomotive in 1930. Although this locomotive differed from the Birmingham & Gloucester examples in several respects, I used the drawing for details of the framing and gab gear.

The tender as drawn is a conglomerate of various Norris examples with outside frames, and is as close as I can get to that shown in Dolby's painting whilst using Norris material. I have not included detail of the boiler-mounted safety valves, believed to have been fitted to these engines, as I do not have sufficient information on

them; the same reason accounts for the lack of sanding apparatus in the diagram, plus the fact that I am unsure when, and if, all the 'B' types received it.

A drawing was published in The Steam Engine by P. R. Hodge in New York in 1840 purporting to be of the Birmingham & Gloucester's Victoria, but since the accuracy was of a 'B' type and Victoria an 'A' type, the accuracy must be called into question. Additionally the Hodge drawing has the buffer planks too low, and shows an engine with a wood-burning type of spark-arresting chimney which the Birmingham & Gloucester locomotives did not have. For many years this drawing was taken as a true representation of a Birmingham & Gloucester Norris engine, but, for the reasons outlined above, I have discounted it and tried to produce as accurate a diagram as possible via other means.

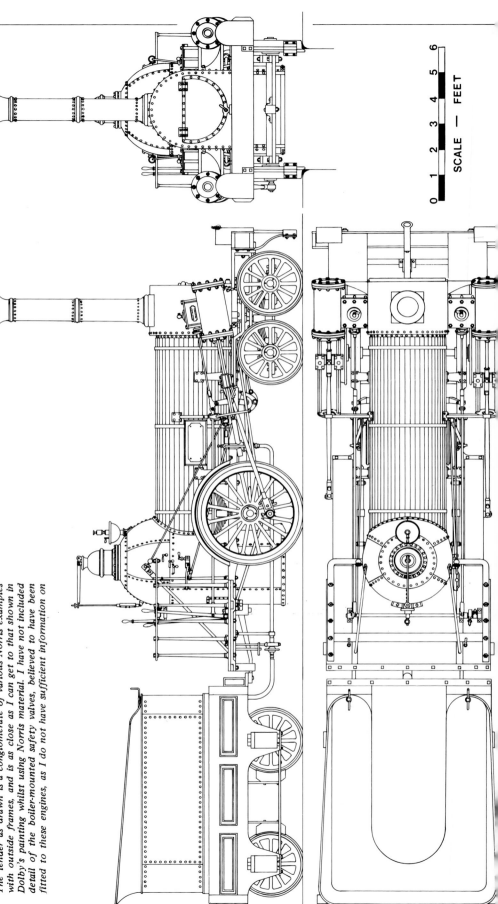

SCALE — FEET

0 1 2 3 4 5 6

That Stephensons could not build an engine of sufficient power is patently untrue as shown by Warren[7], and it is far more likely that English firms declined to tender for locomotives of the same weight and boiler pressure as that advertised for the Norris engines to produce the required performance up the Lickey incline. The only reference in the minutes to refusals of English firms to tender occurs in January 1840 when it is recorded that Bury and Stephensons declined an offer to tender for locomotives 'similar to the Victoria'. For several years prior to 1840, four- and six-coupled engines working at similar boiler pressures to the Norris locomotives (or, at least, their supposed working pressure) had been available from English manufacturers which could have done the job just as well or even better; whether an English-built single-driver locomotive capable of matching the apparent hill-climbing capabilities of the Norris engines was available is, of course, another matter.

Returning to the agreement with Norris, it is sad to say that the stipulated performance on the inclined plane was outside the capabilities of even the largest Norris type, the 'A Extra', at its advertised boiler pressure; it would have required something like a hundred pounds per square inch to achieve it. Even worse was the fact that the larger of the two engines sent was only a smaller 'A' type, the 'A Extra' having, in all probability, not progressed to the stage of actually being built. It is difficult to understand why an 'A Extra' would not have been sent over if one had been available. The trial performance specified for the 'General Line' was just within the capabilities of a 'B' type engine operating at fifty-five pounds per square inch boiler pressure. By this stage, Norris was advertising a range of standard types of locomotive referred to by the type designations used above. A summary of these types is set out in Table 1. The details in the table have been taken from a reproduction of a Norris brochure dated 1st January 1841, with the areas and weights converted from square inches to square feet, and pounds to tons. There was also a 'C' type in the Norris catalogue, smaller even than the 'B' type, but since the Birmingham & Gloucester never purchased any, I have ignored them.

The first engine to be sent for trial arrived in England in March 1839 and, since the Birmingham & Gloucester was still under construction, was sent for trial on the Grand Junction Railway. It was *England*, one of the smaller type 'B' locomotives, and in that may lie a clue as to why Norris or Gwynne had persuaded the Gloucester Committee to try two types. If one of the larger type 'A' engines had been available, why was it not sent over first? The first one did not arrive until the following July, and I think it quite likely that the larger engines, as well as their bigger still 'A Extra' brethren which were the subject of the original agreement, existed only on paper. Whatever the situation, *England* did not achieve her performance criteria. Together with two other type 'B' engines which had arrived in this country, presumably sent over by Norris in the hope of an unsolicited sale, it was fitted at the Grand Junction's insistence with a 'mercurial gauge' and thus operated at more or less the advertised boiler pressure. The trials seem to have been fairly haphazard to start with, and Moorsom recorded all sorts of odd loads and results, but finally he had to admit that expectations were not being reached. By May the directors were worried and a meeting was convened to receive a report from Moorsom as to the situation. Astonishingly, he didn't attend, but sent the report to be read in his absence. By now the directors seemed to have lost some faith in their engineer and made a strongly-worded resolution that he

should carry out further trials, ensuring that the proper loads, boiler pressures, and speeds were adhered to. More tellingly, they resolved that 'An experienced Mechanician be appointed by this Committee to proceed with the Engine on several trips in order to report the results of his observations.'

On 4th June, Moorsom reported that the engine had not met its performance criteria, and the directors resolved to 'decline to receive the said engine'. Norris's agent, Gwynne, must have been pre-warned of this outcome as he had a meeting with the directors the same day, at which he persuaded them that, since he was confident of the larger engine reaching its trial specification for the Lickey incline, they should take the *England*. Not only that, but a week later he managed to get the board's decision to cancel the other six 'B' type engines reversed. In doing so he dropped the price of the engines from £1,750 to £1,525 each. When Hicks later tendered to build copies of 'A' types, their price was only £1,150. What a salesman Gwynne must have been!

In November 1839 what should have been an 'A Extra' arrived at Liverpool but, as we have already seen, it was a smaller 'A' engine. From Gwynne's confidence, prior to its arrival, in the performance of this locomotive I would hazard a guess that he, too, was expecting a larger engine. Named *Victoria*, it was given a trial on the Grand Junction's Bolton inclined plane with loading adjustments intended to simulate the Lickey. Moorsom's heart must have sunk as he reported to the directors on 12th November:

> 'I have to report that an Engine called the *Victoria* intended by Mr. Norris to make the performance stipulated for, on the Bolton inclined plane, has been delivered at Liverpool and brought upon the Bolton line. This Engine, however, is lighter, being only about ten tons in weight, than I should have contemplated under the provisions of the Agreement, but in respect of her manufacture is as perfect as was desired.'

He then went on with the bad news that the engine had failed by fifteen tons to achieve the haulage capacity stipulated. Significantly, he again failed to attend the meeting convened to examine the trial results and sent his report instead.

The directors now found themselves in what could be described as a hole. The opening of the line was fast coming

TABLE 1
NORRIS LOCOMOTIVE TYPES

Type	A Extra	A	B
Cyl. Dia.	12.5in	11.5in	10.5in
Stroke	20in	20in	18in
Boiler length	14ft 6in	13ft	12ft
No. of tubes	97	97	78
Dia. of tubes	2in	2in	2in
Heating surf.	514sq ft	425sq ft	369sq ft
Dr. wheel dia.	4ft	4ft	4ft
Bogie wh. dia.	2ft 6in	2ft 6in	2ft 6in
Weight	13.23 tons	10.76 tons	9.2 tons
Adhesive wt.	9 tons	7.5 tons	5.7 tons

7. J.G.H. Warren *A Century of Locomotive Building 1823–1923* published in 1923. In his work on the history of the first hundred years of Robert Stephenson & Co., Warren gives details of locomotives such as the 0-6-0 produced in 1834 for the Leicester and Swannington Railway which had 16in × 20in cylinders, 4ft 6in diameter wheels, 666sq ft heating surface, and weighed 17 tons, all of which, fairly obviously, was on the coupled wheels.

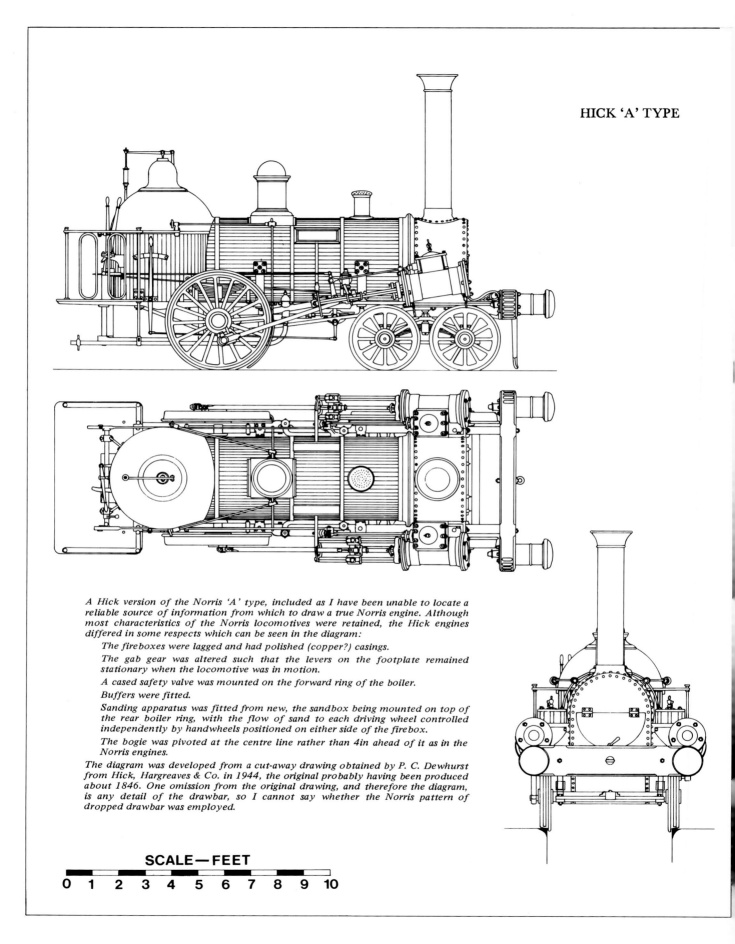

A Hick version of the Norris 'A' type, included as I have been unable to locate a reliable source of information from which to draw a true Norris engine. Although most characteristics of the Norris locomotives were retained, the Hick engines differed in some respects which can be seen in the diagram:

The fireboxes were lagged and had polished (copper?) casings.

The gab gear was altered such that the levers on the footplate remained stationary when the locomotive was in motion.

A cased safety valve was mounted on the forward ring of the boiler.

Buffers were fitted.

Sanding apparatus was fitted from new, the sandbox being mounted on top of the rear boiler ring, with the flow of sand to each driving wheel controlled independently by handwheels positioned on either side of the firebox.

The bogie was pivoted at the centre line rather than 4in ahead of it as in the Norris engines.

The diagram was developed from a cut-away drawing obtained by P. C. Dewhurst from Hick, Hargreaves & Co. in 1944, the original probably having been produced about 1846. One omission from the original drawing, and therefore the diagram, is any detail of the drawbar, so I cannot say whether the Norris pattern of dropped drawbar was employed.

SCALE—FEET

0 1 2 3 4 5 6 7 8 9 10

upon them, they had insufficient motive power on order, and what they did have was, in modern parlance, a busted flush. On 4th April 1840 Moorsom wrote to them again (he seems to have had an aversion to attending in person):

'I should inform you that I have written to Mr Norris to know how soon he could send over two Engines of this class (he refers to the 'A' type) as it is evident no Engine can be made in England in time for your first opening, and I have requested Mr Norris to state his lowest price of delivery.'

Gwynne had promised back in November that he would ensure Norris provided an 'A Extra' if the Birmingham & Gloucester bought the *Victoria* and the six 'B' types. As a concession he allowed that a quarter of the purchase price be reserved until the 'A Extra' was delivered. Once again, the directors gave in to Gwynne's blandishments, probably because they had little option with the opening of the line getting ever nearer, but proved a little more astute in their negotiations than previously and managed to bring the agreed price down to £1,200.

There was still a need for more locomotives, in addition to the Norris ones already delivered or on order, before the line opened, though, and the directors took the surprising step of ordering more Norris types from English 'Respectable Makers'. Altogether they contracted for six from Nasmyth, Gaskell & Company, and three from Benjamin Hick, all based on the Norris 'A' type. Why they would do so in the knowledge that such engines were not up to the Lickey incline is beyond me when more powerful coupled examples were being built by English firms, and Nasmyth and Hick could, presumably, have copied those. As Ahrons, Dewhurst, and others have pointed out, the Leicester & Swannington 0-6-0s of 1834 were more powerful than the Norrises and would have been more suited to hill-climbing at their advertised boiler pressures and adhesion weights. Additionally, three more smaller 'B' types were obtained from Norris with even less haulage capacity. The policy of motive power acquisition by the Gloucester Committee at this time is something I find incomprehensible given the evidence they had before them; even availability versus opening deadlines cannot explain the orders for Nasmyth and Hick. The only thing I can see in their favour was Moorsom's statement in January of 1840 that they were admirable machines for hauling 100-ton loads at 20mph and that, 'I should prefer them to any English Engines with which I am acquainted'. Moorsom seems to have had a fixation with them; one wonders why, and what sort of enquiries would be made into the situation today.

The first 'A Extra' engine finally arrived on 26th May 1840 and was tested on a one mile stretch of track at the bottom end of the Lickey incline itself. Interestingly, the gradient was reported at this time as reaching 1 in 34½ in places. Moorsom's report on 21st July was more optimistic than those following the 'A' and 'B' type trials and includes a couple of interesting points:

'Gentlemen, In reference to my former Reports of the Bank Engine 'Philadelphia' I now beg further to add that having through Mr. Gwynne's assistance fixed a proper Sand Box to the Engine and a Water Tub for wetting the rails, the Engine has been working since Monday the 13th instant with wet rails generally taking advantage of the showers when such occurred and watering the rails at other times. The result has been, that with wet rails or rails partially wet, the Engine works with the Sand Box almost precisely in the same way as with dry rails and without sand.'

This is one of the earliest references I have ever seen to sanding the rails from a moving locomotive as an aid to adhesion, the only other reference as early as this being a mention of similar trials on the Newcastle & Carlisle Railway at about the same time. From the wording of the report, I would assume that it was a system suggested by Gwynne and, therefore, possibly in use already in America. Dewhurst seemed to think that wetting the rails was also used in America as an aid to adhesion, but I read the report as testing the engine's performance in the wet rather than using water for traction. Hot water from the boiler had been used to clean the rails by John Melling, foreman of the workshops on the Liverpool & Manchester Railway, in the 1830s, and the same technique, in addition to sanding of the rails, was employed on two 0-6-0 banking engines designed by Paton in 1844 for the Edinburgh & Glasgow Railway. The reference to 'water tubs' in the report by Moorsom, however, does not seem to indicate the use of hot water. It is possible, of course, that water was used to get the sand to stick to the surface of the rails and prevent it blowing away before it was able to do its job, but I can't see any evidence of 'water tubs' in contemporary drawings or illustrations, and mixing water with the sand before dropping it onto the rails would probably have resulted in blockages. I would be interested in any comments from enginemen on the subject. The sand-boxes were mounted on top of the boiler on the rear ring of the barrel and were either cylindrical or dome-shaped.

Moorsom's report stated that the 'A Extra' had taken 75 tons gross up the incline with a final speed of 10 miles per hour and 55 tons gross at 15 miles per hour. This, of course, would not necessarily represent the performance up the full two miles plus, but he concluded:

'As the latter weight, namely thirty six tons behind the Tender, is more than equal· to your present trains to which a second Engine will be attached, I calculate upon taking the Trains up at the speed of fifteen miles per hour, as soon as the next Engine arrives about mid August.'

THE NORRIS TYPES IN SERVICE

In September 1840, the Lickey incline opened to traffic with just two of the 'A Extra' engines in use, and a third, finally delivered in December just in time for the opening of the line throughout on 17th of that month, on order. The other Norrises were used on the 'General Line', 'A' types normally stationed at Gloucester and 'B's at Birmingham. Trains up the Lickey incline were banked by the 'A Extras' which were kept at Bromsgrove for the purpose. Contemporary accounts speak of 50-wagon trains hauled by three or four locomotives banked by an 'A Extra' but the only illustration I have seen of a Norris engine in action is a painting by Dolby purporting to show *Philadelphia* with eight wagons behind, about the maximum load of one of these engines could cope with unaided. The locomotive in the painting, however, is proportionally inaccurate although most details are in keeping with Norris engines. The forward safety valve does not look like a Norris item and could have been a locally produced one.

The Norris engines were a poor lot in the eyes of the Birmingham & Gloucester Railway's secretary, William Burgess, and two weeks after the Lickey incline was opened he reported that, 'Nearly the whole of the Americans are very much out of order.' The main problems of which he complained were leaking tubes and fireboxes, the former made of copper and the latter of iron in direct opposition to

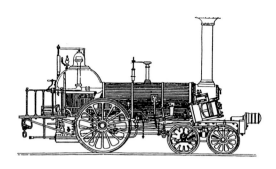

This drawing also purports to show Philadelphia *and is taken from C. E. Stretton's* The development of the Locomotive *(fifth edition) published in 1896. In the original 1893 edition it was referred to simply as 'A bogie engine of improved type'. There are several mistakes in Stretton's work, not least of which is that he records* Philadelphia *in a list of engines having 10½in cylinders and weighing 9½ tons, i.e. 'B' type, whereas it was an 'A Extra'. The drawing does not match what is known of the engine in several respects: firstly, the drawbar is unlike that in any other drawing I have seen of a Norris 4–2–0; secondly, the regulator is of the lateral quadrant type fitted to British-built 'A' types; thirdly, the driving wheels are 4ft 9in in diameter, a size which McConnell's evidence to the Gauge Commissioners stated as applying to 'A' type engines, not to 'A Extras'; the gab gear and valve spindle operating levers are modified and appear to be very similar to the type shown in Lane's illustration of a Nasmyth engine; the locomotive is fitted with buffers unlike Dolby's painting which depicts* Philadelphia *without buffers even after the sanding apparatus was fitted. In fact, the drawing is closer to the Nasmyth 'A' type than anything else, except for the slidebars and the iron boiler cladding shown in Lane's illustration. It almost looks as though it represents a Nasmyth engine which received Norris pattern slidebars and had the iron*

boiler cladding removed when it was rebuilt with 4ft 9in wheels, although the size and height of the cylinders do not match up with the Lane illustration either. Since there are so many anomalies, I have not attempted to produce a diagram from this drawing.
AUTHOR'S COLLECTION

the normal English practice.[8] The tubeplates also were criticised for being too thin. The 'A Extras' had an intermediate support plate halfway down the boiler to prevent the tubes from sagging. Several locomotives had been retubed and copper fireboxes were being fitted in place of the originals. Even the great Norris supporter Moorsom was prompted to write to Philadelphia complaining, albeit mildly, of the poor fit and finish of recent engines, in particular one of the 'B' types, *Gloucester*, which had been taken out of service for repairs in June 1840 less than a month after arriving. He reported to Norris that:

[The engine] 'Became very leaky after a few trips and also allowed much radiation from the smokebox, the effects of which have been well evinced in her consumption of coke and water which has been during our trial runs a large proportion in excess of her three comrades.'

In April 1841 the third 'A Extra', *Boston*, blew back through the firebox whilst ascending the Lickey incline and killed the Birmingham & Gloucester's locomotive superintendent, Mr. Crewze. The reason for the blow-back was given as a modification to the boiler, where a tube had been replaced by a washout plug, failing. *Boston* was repaired but a new locomotive superintendent was needed. It would seem that Crewze was pretty unreliable and ineffective anyway, being a well-known drunkard, and didn't seem to have much to contribute to locomotive matters beyond everyday running. Dewhurst records that Moorsom's candidate since 1840 for the position of locomotive superintendent had been Gwynne, but for unknown reasons this was not to be and the Board appointed James McConnell, later to become locomotive superintendent on the southern division of the London & North Western Railway; at once he began to make changes. As well as fireboxes being replaced, McConnell changed the layout of the firebars which, as originally built, were spaced to suit wood burning and dropped a lot of coke through the gaps, and made adjustments to the valves. He also altered wheel diameters on some engines to 5ft, 4ft 9in, and 4ft 3in. Unlike Moorsom, who seems to have been almost blind to shortcomings of the Norris engines, McConnell was only too conscious of their main failings. Consequently, he asked for, and in 1842 received, permission to undertake extensive rebuilds of the 'A Extras'. In February of 1842 he rebuilt *Philadelphia* as a saddle tank so as to increase the adhesion weight, insulated the boiler with

felt, and fitted coke bunkers on either side. These were not fixed in place but were removable. Steam from both the exhaust and the safety valves was used to preheat the water in the 400-gallon tank and the rebuild resulted in a drop in coke consumption from 92 pounds per mile to 43 pounds per mile.

At the same time as this was going on, and incredibly only a few months before the first of the Norris engines was sold, the Birmingham & Gloucester Railway bought another two 'A Extras'. They appear to have been another of Norris's speculative exports and the first mention of them was in the Birmingham & Gloucester minutes of March 1842 when it was recorded that:

'Two engines of Mr. Norris similar to the Company's Bank Engines now lying at Liverpool, and which Mr. Norris had through the medium of the Engineer in Chief offered for sale to this Company. Ordered — That the Engineer in Chief be informed that the Committee will give two thousand pounds for the two Engines.'

It seems surprising, to say the least, that Norris was still sending locomotives over from America on a speculative basis and that Moorsom was recommending their purchase, but at least they were 'A Extras' and therefore of use on the Lickey incline. The initial response by Gwynne to the company's offer of £2,000 for the two locomotives was to start haggling, but the Committee was firm and, since there were presumably no other takers, they duly arrived at Bromsgrove in May. Considering that the price of the first 'B' types to be delivered was £1,750 each, it is notable that in a few years the 'A Extras' were only considered to be worth just over half that amount.

The other two original 'A Extras' and the two 1842 acquisitions were converted by McConnell in the same year to

---

8. Normal British practice was to use copper fireboxes and brass tubes. Later on, several British designers tried using steel fireboxes but, unlike American and Canadian experience, they were never very successful, due mainly to cracking and leakages. A commonly quoted reason for this was the different carbon content of British versus American steel, but some Canadian locomotives used British steel successfully, and it was more probably differences in firebox design which caused the trouble. Tubes of solid drawn copper, as well as other alloys, were later used by some designers.

saddle tanks following his successful experiment with *Philadelphia*. According to Ahrons, one or two of the Norrises were converted to 4-4-os sometime between 1846 and 1850, but no other reference to such an experiment seems to exist, nor does Ahrons comment on the success of these supposed rebuilds or their subsequent history, and I would treat the story as apocryphal.

By this time the unsuitability of the 'B' type engines had dawned on the Birmingham & Gloucester directors (or, at least, they had decided to do something about them) and *England* and *Pivot* were put up for sale. McConnell was also authorised in 1843 to dispose of another four on the best terms he could get. Unfortunately for the Birmingham & Gloucester, the terms weren't very good — no one wanted them. By the end of 1843 only two had been sold and it wasn't until 1845 that any more were disposed of. The last one to go, ironically *Pivot*, one of the first to be put up for sale, was finally withdrawn by the Midland Railway in 1855. In his evidence to the Gauge Commissioners in August 1845, McConnell gave details of the Norris type locomotive stock of the Birmingham & Gloucester Railway as follows:

Three with 12.5in × 20in cylinders and 4ft wheels ('A Extra').
Two with 12.5in × 20in cylinders and 4ft 3in wheels ('A Extra').
Twelve with 11.5in × 20in cylinders and 4ft 3in, 4ft 9in, and 5ft wheels ('A' type).
Seven with 10.5in × 18in cylinders and 4ft 3in wheels ('B' type).

From this return it would seem that McConnell had been experimenting with various wheel diameters on all the Norris types and that four feet three inches was a preferred option. Unfortunately, I have been unable to trace any records of these rebuildings or perceived results of the changed wheel diameters.

In their brief lives on the Birmingham & Gloucester, the 'A' types generally hauled trains of 45 tons gross and attained, if Moorsom is to be believed, speeds of 34 miles per hour, occasionally reaching 38 miles per hour, on the level, and 24 or 25 miles per hour up 1 in 300 gradients. They were used on all trains except the fast mails.

Of the 'A' type engines, the first to go were, surprisingly, the Nasmyth and Hick-built copies, and all three of the Norris products were taken into Midland Railway stock, finally being disposed of by 1855. The rebuilt 'A Extras' were also taken into Midland stock and the last of them was recorded by F.S. Williams as being used on the Tewkesbury branch. I should say that Williams records the last of the Norris engines on the Tewkesbury branch in 1856 and I am assuming it was one of the rebuilt 'A Extras'. From the withdrawal dates I assume it would either have been *Philadelphia* or *Boston*, by then Midland Railway numbers 271 and 272. Only one 'B' type, the aforementioned *Pivot*, survived to become Midland property.

Lest I give too biased a picture of the Norris engines, I should add that there were some good things said about them by contemporary observers. David Joy wrote of one of them:

'The little thing could pull, but she was odd, lots of cast iron in her, even the cross head pins were cast iron.'[9]

They were also maligned by being blamed for a fatal explosion in 1840 when the foreman of locomotives at Bromsgrove, Joseph Rutherford, and one of the bank engine drivers, Thomas Scaife, were killed. In fact the engine concerned was a tank locomotive designed by a Dr Church, and built by Henshaw, which was undergoing trials on the Birmingham & Gloucester Railway. Henshaw's son was firing it when Rutherford and Scaife went over to take a look at the engine, unluckily just at the time it decided to explode; Henshaw junior was badly scalded but survived. The name of the engine was, appropriately enough, *Surprise*. The mason carving the dead men's headstones, however, chose to portray Norris locomotives and so they got the blame for Scaife and Rutherford's deaths for many years.[10] There doesn't seem to be any record of a boiler explosion in a Norris engine on the Birmingham & Gloucester apart from the blow-back previously described, and that wasn't the fault of Norris's design. They were also reputedly good steamers and the 'A Extras' were recorded as blowing-off from the safety valves all the way up the Lickey incline when banking trains.

CONSTRUCTION

The Norris engines as built had relatively light bar frames with the cylinders attached both to the frames and the boiler. Cylinders were reported by contemporary observers as being copper clad and not to have any packing. According to the *Mining Journal* of 1st June 1839:

'The steam tight joints formed simply by bringing into contact the metallic surfaces, the workmanship of which is so true as entirely to supersede the necessity for packing of any kind.'

Tie bars ran from the driving wheel horns forward to the frames, just above the rear truck wheels, and aft to the underside of the footplate, just under the rear of the firebox. The firebox and motion brackets were stayed to the footplate and boiler respectively, and the slidebars were of diamond cross section mounted alongside each other. Ahrons gives them as being made of cast iron, but Dewhurst doubts this statement, estimating that cast iron would not have been up to the strains imposed; as I have seen no reports of weakness in the slide bars I tend towards the latter view. Connecting rods were round section wrought iron attached to the driving wheels via brass cottered big ends between the spokes and to fabricated crossheads. The boiler barrel was lagged with pine slats held in place by brass bands, but the fireboxes were unlagged. Firebox design differed between the types, the 'A Extras' having a haycock firebox of internal diameter greater than that of the boiler, the 'A's the same diameter, and the 'B's smaller. They were stayed to the frames by rods running from near the top to the rear of the running plate. The tops of the fireboxes were hemispherical, surmounted by brass dome-like structures, on top of which were the spring balance safety valves. The piston valves were actuated by simple four eccentric gab gear operated by a lever alongside the firebox, and the regulator operated in a fore and aft sense rather than on a lateral quadrant. The boilers were supplied by crosshead-driven pumps mounted on the frames just ahead of the driving wheels. The cylinder cocks could not be operated from the running plate, but only from taps alongside the cylinders. The open footplates

9. Extracts from the diaries of David Joy were published in *Railway Magazine* in July 1908. Joy worked at E.B. Wilson's and was responsible for the design of the highly successful 'Jenny Lind' class of 2-2-2 which the Midland Railway bought in 1848.

10. There have been many inaccurate stories printed about this accident, suggesting that Scaife and Rutherford were the crew of the unfortunate locomotive and/or that the engine was named *Eclipse*.

The graves at Bromsgrove of Thomas Scaife and Joseph Rutherford, killed by a boiler explosion at Bromsgrove on 10th November 1840 which has long been erroneously blamed on a Norris locomotive. The verse on Scaife's headstone is interesting and reads:

'My engine now is cold and still
No water does my boiler fill
My coke affords its flame no more
My days of usefulness are o'er
My wheels deny their noted speed
No more my guiding hands they heed
My whistle too has lost its tone
Its shrill and thrilling sound is gone
My valves are now thrown open wide
My flanges all refuse to guide
My clacks alas though once so strong
Refuse to aid the busy throng
No more I feel each urging breath
My steam is now condensed in death
Life's railway's o'er, each station's past
In death I'm stopp'd and rest at last
Farewell dear friends and cease to weep
In Christ I'm safe in Him I sleep.'

Also of note is the fact that whereas Rutherford's gravestone describes him as an engineer to the Birmingham and Gloucester Railway, that of Scaife refers to the Birmingham and Worcester Railway.
P. KIBBLE

A close-up of the locomotive carved on Thomas Scaife's gravestone, erected in 1842. It seems to be a fairly accurate representation of a Norris engine and matches closely what is known about the Norris 'B' type, though it is, of course, possible that it was an 'A' type. If so, it was a Norris-built one rather than a Nasmyth or Hick copy and was fitted with a sand box on the rear ring of the boiler barrel.
P. KIBBLE

The locomotive depicted on Joseph Rutherford's headstone was carved earlier than Scaife's in 1841 and shows a different engine. It was a bigger machine than the one carved in 1842, the rear stays from the driving wheel horns reached to the back of the platform rather than the rear of the firebox, and the sand box was mounted further forward on the boiler barrel. Assuming that the mason carved accurate representations of the locomotives he saw, I would hazard a guess that the one shown in this illustration was an 'A Extra', although it lacks the extra forward safety valve and raised buffer blocks of Philadelphia as depicted in the painting by Dolby.
P. KIBBLE

were fitted with handrails mounted on stanchions. Wheels were cast iron, 12 spokes on the trucks and 16 spokes on the drivers, with chilled cast-iron tyres.

The front trucks had no sidewards movement, being merely pivoted about a point some four inches ahead of the centre, but the pivot took none of the load; instead, extensions of the spring buckles bore on slides attached to the undersides of the frames. Thus the weight was taken at the mid point of the truck, and the forward position of the pivot tended to counteract sway. British locomotive engineers considered bogies to be unsafe, particularly at high speed, in the 1830s and 1840s. At this time they did not have side control nor systems to prevent twisting of the bogie across the tracks, and since the wheelbases on bogies tended to be less than the track gauge (in the case of the Norris engines only three feet), there would always be a degree of instability in this sense. It is possible, therefore, that the Norris system of pivoting the bogie ahead of its mid point could have been an early attempt to overcome this problem, although I have never seen it suggested or commented on by other writers. Certainly there does not seem to have been any problem with stability of the engines. If this was the case, though, it seems strange that the Nasmyth and Hick copies of the Norris design did not include this feature unless, of course, they did not realise why it had been incorporated. I have not seen any references, however, to comparative stabilities of the American-built and English-constructed locomotives.

As previously stated, the tubes were copper when the locomotives were delivered, and the fireboxes of iron. The tube plates initially were in two pieces but this made them prone to leaks, and after the 'Gloucester' fiasco it was reported by Burgess that:

> 'Another engine of Victoria Class ['A' type, probably *Baltimore*] has now landed at Liverpool, and that the tube plate of this engine is in one piece, so that I hope we shall avoid part of the present evil.'

Even so, the firebox plates were said to be very thin. In some cases the original fireboxes only lasted eight months before being replaced by copper ones. Smokeboxes were riveted up from iron plates and had semi-circular doors in the lower half secured by a single handle. Several writers, including Moorsom, complained of them leaking. The locomotives did not have real buffers, merely upward extensions of the front beams with wooden blocks.

THE BRITISH COPIES

Although strictly outside our present scope, as they weren't actually American in origin, the Nasmyth and Hick engines differed in some respects from the true Norrises. The Hick examples had copper fireboxes from new, extra lock-up safety valves on the firebox, centrally-mounted truck pivots, buffers, and different chimneys. The boilers contained 96 brass tubes, each two inches

A wash illustration made in 1849 by E. T. Lane of a Nasmyth version of an 'A' type Norris 4—2—0. As some of the details and dimensions seem a little suspect, I have not produced a diagram from the illustration. It is interesting, however, in that it shows the laterally-operated quadrant regulator, modified gab gear, iron-clad boiler, and four rectangular section slidebars of the Nasmyth engines. Strangely, though, it does not show any sanding apparatus nor boiler-mounted safety valve, both of which are reported by contemporary accounts to have been fitted to the locomotives from new.
AUTHOR'S COLLECTION

diameter. The reversing levers on the running plate remained stationary when the engine was running, rather than oscillating as on the Norris ones, due to an improvement in the gab gear. Nasmyth altered the slidebar arrangement to four wrought-iron ones, and fitted a laterally operated regulator and wrought-iron tyres. Nasmyth engines also had improved gab motion, centrally pivoted trucks, buffers, and, if a contemporary illustration is accurate, iron plates around the wood boiler lagging. The chimneys were different from either the Norris or Hick types. The Hick engines seem to have been satisfactory machines, but the Nasmyth versions came in for adverse comment, the first one to be delivered, *Deptford*, being found to be defective on arrival. Another one, *Droitwich*, took several days to make the journey from Patricroft to Bromsgrove due to repeated failures en route.

No other British railway made extensive use of the Norris locomotives; indeed, the only others who made any use of them at all were the Taff Vale, who bought two Birmingham & Gloucester Railway cast-offs, and the Aberdare Railway which bought one. The most successful variant was the McConnell rebuilt 'A Extra' which, in its saddle tank form and with modifications to curb its appetite for coke and thirst for water, performed well on the Lickey incline until McConnell's own specially designed Lickey banking engine *Great Britain* appeared in 1845 and began taking 135-ton trains up to Blackwell. The Norris type influenced locomotive design on the continent of Europe and was widespread for some years, but, like its contemporary, the Bury type, in Britain it was destined only for a brief and limited career.

The Norris works continued to produce locomotives until 1866 when it closed, and the site lay derelict until it was taken over by Baldwins in 1873, which connects neatly with the Midland's second period of American Locomotive operation, more of which in Chapter Two.

CONCLUSION

A summary of the Norris engines is given in Table 2. Where I have found conflicting dates or details, I have given the alternatives; where data is derived from a non-contemporary source and it not supported elsewhere, I have indicated such by a question mark. Note that until McConnell became locomotive superintendent, Birmingham & Gloucester Railway locomotives carried only names and did not have numbers. Also, although the engines which survived into Midland Railway ownership were operated by that company from August 1846, they were not numbered into Midland stock until February 1847. Thus some of them never received Midland Railway numbers, although they were, for a short time, Midland property.

Details of the Nasmyth & Hick-built engines are given in Table 3. Notes, alternative dates, and unsupported details are indicated as in Table 2.

Apart from numbers 10 and 11, which were sold to the Taff Vale Railway, and number 8 which went to the Aberdare Railway, all the locomotives sold went to contractors or private lines.

So, the first experience of the Midland Railway, or one of its constituents, of American locomotives was not an outstandingly happy one. The engines were bought to fulfil a particular job at a time when, for various reasons, the directors of the Birmingham & Gloucester Railway felt themselves unable to go elsewhere to meet their requirements. The locomotives were the subject of controversy and mixed comment in their lifetimes, and none of them lasted a long time. In these respects, as we shall see shortly in Chapter Two, there were uncanny parallels with events sixty years later when the Midland Railway proper turned to locomotive builders across the Atlantic.

TABLE 2
NORRIS LOCOMOTIVES ON THE BIRMINGHAM & GLOUCESTER

Name	Date	B&G No.	MR No.	Remarks
		NORRIS 'B' TYPE		
England	3/39	5	–	Sold by end of 1842
Atlantic	3/39	7	–	Sold 7/46
Columbia	3/39	8	–	Sold 1/46
Moseley	2/40	16	–	Sold 12(?)/43
Pivot	7/40	17	178	Renumbered 107 6/52. Withdrawn 1855
Birmingham	6/40	9	–	Sold 4/46
Gloucester	6/40	10	–	Sold 9 or 10/45
W.S.Moorsom	6 or 7/40	11	–	Sold 8 or 9/45
Washington	8(?)/40	12	–	Sold 11/46
		NORRIS 'A' TYPE		
Victoria	11/39	6	277	Renumbered 106 6/52 Withdrawn 1855
Baltimore	9/40	15	281	Sold 1852
President	12/40	20	282	Sold 1852
		NORRIS 'A EXTRA'		
Philadelphia	5/40	13	271	Renumbered 113 6/52 Withdrawn 6/56
Boston	8/40	14	272	Renumbered 114 6/52 Withdrawn 6/56
Wm. Gwynne	12/40	21	273	Sold 5/52
Niagara	5/42	31	274	Sold 5/51
New York	5/42	32	275	Renumbered 104 6/52 Sold 4/55

N.B. All 'A Extra' type engines converted to saddle tanks in 1842.

TABLE 3
NORRIS TYPE ENGINES BUILT BY BRITISH FIRMS

Name	Date	B&G No.	MR No.	Remarks
		NASMYTH 'A' TYPE		
Defford	11/40	19	280	Renumbered 109 6/52 Scrapped 12(?)/55
Ashchurch	1/41	26	–	Sold 8/46
Droitwich	1/41	27	–	Scrapped(?) 1847(?)
Pershore	6/41	28	–	Sold 4/46
Upton	7/41	29	–	Sold(?) 3/47
Lifford	8/41	30	–	Sold 10/47
		HICK 'A' TYPE		
Bredon	10/40	18	279	Renumbered 108 6/52 Scrapped 12/55
Spetchley	1/41	24	–	Sold 3/47
Eckington	1/41	25	–	Sold 10/46

CHAPTER TWO
THE MOGULS

THE only 2-6-0s ever owned by the Midland Railway were of American origin too. They were also among the relatively few outside-cylindered Midland engines, the only other classes of Midland Railway locomotives with this feature, apart from one-offs, being the Metro tanks, Johnson and Deeley/Fowler 3-cylinder 4-4-0 compounds and the Deeley 0-4-0 tanks. As mentioned before, there were links between the previous Norris engines and the later American machines. In 1873 Baldwins, by then the biggest locomotive builder in the United States, had acquired the defunct Bush Hill works of the Norris company in Philadelphia. Prior to 1862 Norris Brothers had been the biggest locomotive builders in the United States, pre-eminent in both output and reputation, but, after the American Civil War, things began to slide. The workmanship of their engines came in for major criticism, orders fell away, and after a relatively short time the firm slipped from its position of supremacy into oblivion. The mantle of leadership in American locomotive practice passed to Baldwins. Baldwins' works were just across the street from those of Norris and it must have been satisfying to occupy the former premises of their rivals. It was from Baldwins that the first of the Midland Railway moguls came in 1899, joined shortly after by similar engines from the Schenectady works in New York State. A somewhat tenuous link also existed between Norris and Schenectady as one of the Norris brothers, Edward, had set up a locomotive works there in 1847, but within four years had gone out of business. To be absolutely accurate, the Philadelphia establishment by this time was actually the Baldwin works of Burnham, Williams & Company, but it was generally known simply as Baldwins. The Schenectady works later became the nucleus of Alco, one of the giants of American locomotive building.

THE MOGUL TYPE

The 2-6-0 wheel arrangement had been pioneered by both Norris and Baldwin at Philadelphia in 1852 and 1853; initially, however, the type was not a roaring success and after a short time mogul production ceased and was not resumed for another ten years or so. The reason for the type's unpopularity was the fact that the leading axle in both versions was fixed to the frames ahead of the leading coupled axle but behind the outside cylinders. As with the Stephenson long-boiler type, therefore, the design was somewhat unstable due to the long overhang of the considerable mass of smokebox and cylinders ahead of the leading wheels. These wheels, being fixed in the frames, were not able to do the job of a proper pony truck and guide the front end of the locomotive into curves. Outside cylinders, by this stage standard American practice due to a dislike of crank axles, added to the problem because of the oscillatory motion of the pistons and the greater lateral moment when compared with cylinders mounted between the frames. The outcome was that the early moguls suffered frequent derailments and, not surprisingly, their popularity evaporated. In May of 1857 the Bissell truck was patented in Britain and about three years later was employed on locomotives being exported to several countries including the United States. In 1864 Baldwins designed a mogul with the leading wheels mounted in a Bissell truck and ahead of the outside cylinders. Because the truck controlled the sideways motion of the engine ahead of

the front end mass, the predisposition to oscillate was curbed, and the same side control guided the leading coupled wheels into curves, so that the tendency to derail was overcome. These problems were more pronounced in the United States because the track, generally lighter than that used in Britain, often spiked directly to the sleepers, and frequently on a less carefully laid track bed, gave locomotives a rougher ride and highlighted stability or springing shortcomings very quickly.

The Bissell truck was constructed as follows: Two lugs projected upwards from inside the frames and above the axle boxes, through which longitudinal pins passed, carrying the upper ends of a pair of swing links. The lower ends of the links were pinned to projections from a centre casting, which was carried in a central socket fixed to a cross member between the locomotive frames. Thus the whole truck could move laterally with reference to the frames by literally swinging on the links. To prevent the wheels twisting about the central pin, the truck frames were extended rearwards in a vee shape to another pin further back between the main frames. The axleboxes were supported by beams which extended down either side of them to bearers for coil springs, the other ends of which were attached to the truck frames. Coil springs were initially used as they are faster acting than leaf springs, considered to be a distinct advantage for leading wheels used to guide a locomotive into curves. Later versions of the Bissell truck used volute springs which are even better than coils.

The claimed advantages of swing link or Bissell trucks were twofold: first of all the action of the links was smoother than slides and springs and had less inherent friction, thus they lessened the strains on the leading wheel flanges when going into a curve; secondly, if the links were arranged so that their lower ends were closer to the locomotive centre-line than the upper ends, the action when the truck moved to one side would tend to put more weight on the outer wheel and so enable the inner one to slip more easily along the shorter path of the inside rail. Bissell trucks were normally equalised with the leading coupled wheels of the locomotive.

THE MIDLAND MOGULS

Once the Bissell truck 2-6-0 was in being, it soon became popular for all sorts of applications in America, and one of the great workhorse locomotive types was evolved. Most American manufacturers produced standard moguls.

In Britain, though, the overwhelming majority of goods engines were inside-cylindered 0-6-0s. The problems of crank axle manufacture and breakages had been overcome sufficiently well for most locomotive designers to use inside-cylindered types. This meant that they could keep the cylinders nice and warm underneath the smokebox and close enough to the longitudinal centreline to minimise any tendency to a swaying motion because of the oscillation of the pistons. Therefore the complications of pony trucks and such were not necessary, certainly for locomotives which would spend their time plodding along at slow speed, and the standard British goods engine type was seen in vast numbers on most British railways. The Midland Railway was firmly wedded to the genre and had built virtually nothing else for its goods traffic since the 1860s. They came in two basic guises, the Kirtley/Johnson outside-framed engines with different running plate profiles, and the pure Johnson inside-framed locomotives, but they were pretty much the same in essentials. Indeed, apart from the 2-8-0s later built for the Somerset & Dorset Joint Railway, the Midland Railway continued for the rest of its existence with the inside-cylindered 0-6-0 as the standard goods engine. At this stage I shall resist the temptation to canter off into the aesthetics of Johnson's rebuilds of Kirtley goods engines — I will leave it to my good friend Jack Braithwaite to wax lyrical on the subject in his contributions to *Midland Record* — but I do think they were beautiful little machines.

Back to American moguls. So how did the Midland Railway come to purchase American 2-6-0 goods engines with outside cylinders instead of simply adding to the bat-

talions of inside-cylindered 0-6-0s from either Derby works or outside British contractors? The move was anathema to many people in this country and provoked a storm of criticism, not just among railway cognoscenti and in the railway press, but in national newspapers. It has to be remembered that all this happened long before any 'special relationship' existed between Great Britain and the United States. In fact, there was antipathy to America and all its doings among a large proportion of the British populace prior to the First World War, and it was in such an atmosphere that the Midland Railway had taken its, therefore controversial, decision to buy locomotives from across the Atlantic. *Railway Magazine* was particularly scathing and began a campaign which it continued long after the arrival of the locomotives, as we shall see later. To quote from Charles Jones, assistant

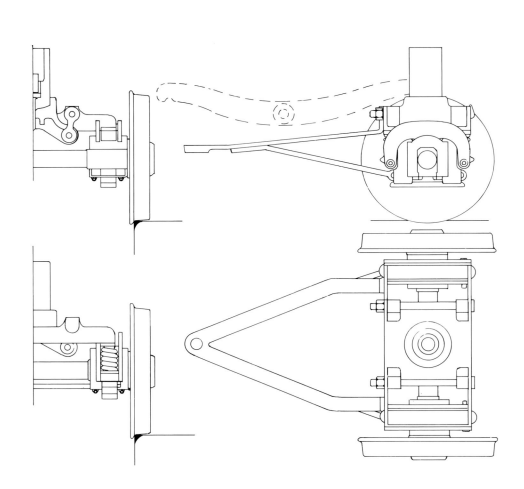

A diagram of a typical Bissell truck. The cross-section in the upper left drawing shows the swing links suspending the truck frame from the centre casting, whilst the other front view and the side view show the coil springs supporting the axleboxes. The compensating lever normally used with Bissell trucks is also shown dotted on the side view. Note the relationship between the pivot points for the swing links on the truck frames and the centre casting. As described in the text, this arrangement put more weight on the inside wheel when the locomotive was rounding a curve.

locomotive superintendent of the Midland Railway under Johnson, in September 1899 (with my comment in brackets):[11]

'The announcement that a large order for locomotives had been placed in America by one of the leading English railway companies was somewhat startling. Many Englishmen regard it as an unpardonable intrusion for foreign engines to be brought to Great Britain, the birthplace of locomotives and of Stephenson, the Father of them. [No xenophobia there then!] If the directors of the Midland Railway Company could have avoided it, they certainly would not have wounded the susceptibilities of these people, but exceptional circumstances rendered it necessary.'

At the annual Midland Railway shareholders meeting at Derby on 17th February 1899, the Chairman of the company, Sir Ernest Paget, felt it necessary to make a statement to the shareholders present on the matter of purchasing locomotives from America, and it is worth studying his words for an explanation for the apparently deeply unpopular move by the Midland board.

'You will no doubt have seen that we have been purchasing some engines in America. As this is a new departure, some explanation will be interesting to you.' [Either he discounted the Norris locomotives as having been bought by the Birmingham & Gloucester rather than the Midland Railway proper or he didn't know about them.] We should very much prefer to purchase home made goods, whether it be engines or anything else, if it were possible. The question of cost did not enter at all into our calculations when we asked for tenders for engines from over the water. Our train mileage has been increasing very rapidly of late years — the train mileage in this last year of 1898 increased by more than two million miles. Our locomotive superintendent, Mr Johnson, has for some years been impressing upon us that we work our engines too hard. [An interesting comment in view of the Midland's reputation for working its engines lightly and employing pilot engines at the drop of a hat.] I believe that if he had his way he would have about seventy five per cent of them in steam; as it is, ninety per cent is much nearer the number, so that you will see that there is no margin there.

'We have at present one hundred and seventy engines ordered in England. The orders commenced in December 1897. The first engines were to be delivered in July 1898, at so many per month, and if that delivery had gone on properly we should now have forty eight of those engines. We have not received one. The last order we gave, in December 1898, was for twenty engines at a very large cost, and we could not get even the promise of one engine for fifteen months, and that order will not be completed until May 1900.

'Now, gentlemen, engines are a necessity to us; we must have them. Therefore we determined to send for tenders from two firms in America, one the Baldwin and the other Schenectady Works, and we received offers from them. In one instance the delivery was to be in ten weeks from the time that they received all the drawings and other data, and, in the other case, shipment from America in four months — ten engines from each — so that you see while we cannot get an engine promised in England in fifteen months, we can get twenty engines from America in four. I do not think we require further justification, and, thinking so, we have doubled the order to one of these firms.

'The engines we have ordered will be of what is known in America as the Mogul type. They will be essentially American engines, but with certain modifications and alterations which our locomotive superintendent has thought necessary. These engines are of the same power as our own, and it will be very interesting to us, and to all Englishmen, I think, to see them running side by side with our own.'

It was actually intended to order another ten locomotives from the Schenectady works, thus making forty

altogether equally divided between the two manufacturers, but, for reasons I have been unable to determine, the final ten were ordered from Baldwins. Possibly the shorter delivery time from the Philadelphia works was the deciding factor, and so the final order was for thirty engines from Baldwins and ten from Schenectady.

What was not explained in Sir Ernest Paget's statement was the reason for such a backlog in deliveries from British manufacturers. Derby Works was not of sufficient capacity to produce enough new engines by itself for the Midland Railway's needs (in fact, it was never able to meet the company's total requirement) and the increase in train mileage in the 1890s referred to by the chairman had made the situation worse, so the company's reliance on outside contractors had made them vulnerable to any delivery hold-ups from those firms. A large proportion of the reported mileage increase had been in mineral traffic, and there was a pressing need for goods engines, but the Midland was not the only British railway to be faced with the same increased demands on its resources, and many other railways were ordering large numbers of locomotives from British builders. Manufacturers' order books were full to capacity when a crippling series of strikes in the engineering industry reduced even further their ability to meet the demand. The flow of new locomotives dried up and the untenable situation of which Sir Ernest Paget spoke left the Midland and other railway companies with little option but to go elsewhere to meet their requirements. Baldwins certainly did well out of British firms' misfortunes as they received orders from two other British railway companies for the same type of locomotive as that built for the Midland — The Great Northern Railway ordered twenty and the Great Central Railway the same number. It has been said by some writers in the past that the decisions of these companies to go to Baldwins for locomotives was prompted by seeing the Midland examples in service, but the deliveries commenced so soon after the Midland Railway's engines arrived in this country that I very much doubt this was the case. Rather I suspect that a simple matter of availability of suitable engines in as short a time as possible, having waited so long in vain for the situation at home to improve, was the deciding factor.

The question was still asked, however, as to why American manufacturers could supply engines so quickly compared with British firms, even without the effects of a major strike to delay things. A contemporary article in Cassier's magazine suggested that the willingness of American employees to work flexible shifts, so that the factories could operate round the clock when demand was great, constituted part of the reason. The article said that factory workers in the United States were prepared to work harder and thus earned more money than their British counterparts. I know insufficient about late Victorian working practices and conditions in the United Kingdom, let alone America, not to mention comparative costs and standards of living, to make a subjective assessment of this remark, so I will leave it at face value. However, I think it possible that more than a little hint of sour grapes may have been present given the amount of bad feeling generated by the engineering strike. What is notable is that the engines ordered were not Johnson's own designs, but, to all practical purposes, the builders' standard products. American firms had a range of stan-

---

11. Jones was writing in *Cassier's Magazine* of September 1899.

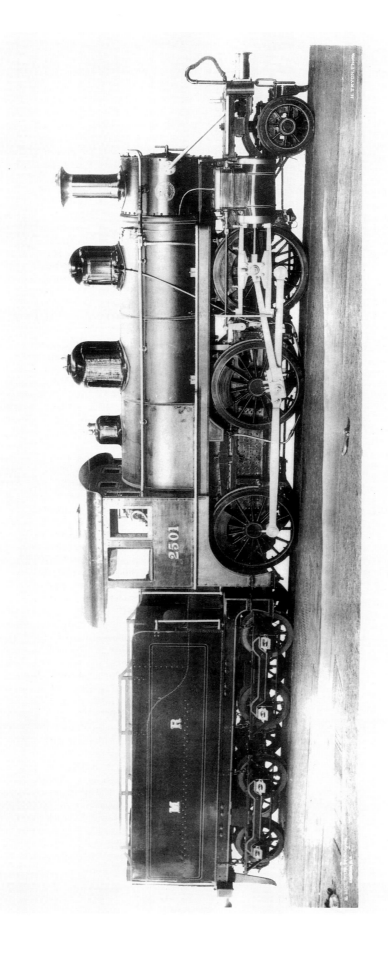

A Baldwin official photograph of the first Midland Railway 2–6–0 from the collection of the late C. E. Stretton. The boiler fittings are shown highly polished and the chimney copper capped. This feature is not referred to in any document I have seen and, as stated in the text, I would imagine it may have been a special addition for publicity purposes, although some contemporary correspondence refers to at least one Baldwin running in Britain so equipped. The open construction of the front frame extensions can be easily seen.
NATIONAL RAILWAY MUSEUM

dard designs available of whichever wheel arrangement and power output the customer wanted; in fact, we saw earlier that even in 1841 Norris had produced a catalogue of four standard engines. Thus the jigs and patterns needed to build a locomotive already existed if it was a standard product, as possibly did many of the components and fittings. Had the Midland Railway's order been for a design other than one of the American 'off the shelf' models, I would surmise that considerably longer than ten weeks would have been required to produce ten engines. It so happened that the standard moguls fitted the bill as far as the Midland, Great Northern, and Great Central railways were concerned. The drawings and data referred to by Sir Ernest Paget, and supplied by the Derby drawing office to the American firms, were for relatively minor alterations. The changes from the standard Baldwin and Schenectady designs specified by Johnson and the Derby drawing office were as follows:

> Fireboxes were to be made of copper, in line with normal British practice, rather than the steel versions preferred in America.
> The wheels were to be of Midland design. This may have been the case, but the driving wheel bosses certainly didn't look like normal Midland ones, being lighter than those usually associated with Johnson's designs.
> Standard Midland buffing and drawgear was to be fitted in place of buckeye couplers.
> Steam sanding equipment was to be of standard Midland pattern.
> Midland pattern combined steam and vacuum brakes were to be fitted.
> Injectors and water gauges were to be of Midland design.
> Johnson pattern smokebox doors were to be fitted.
> The tenders were to have the addition of Johnson pattern coal rails.

In the case of the Baldwin locomotives, that was about the extent of the modifications applied, but the Schenectady 2-6-0s were much more Anglicised, as we shall see later. It would seem, however, that these more extensive changes to the New York engines were not through any insistence on the part of S.W. Johnson or the Midland Railway board, rather, I would imagine, they represented an attempt by the Schenectady people to please the customer. The only major stipulation by the Midland Railway to either of the American manufacturers was that the engines supplied should be equal in power to the standard Midland goods engine, which proviso their 'off the shelf' mogul types satisfied.

Another reason given for rapid fulfilment of orders, and one which I daresay holds some merit, was that American manufacturers did not build locomotives as 'showy' as their British counterparts, and that:

> 'They are satisfied with what is good enough, and do not expend time and labour on more highly finished workmanship that they think is requisite for all practical purposes.'[12]

This could, I suppose, be translated as their being 'cheap and cheerful' and makes interesting reading in view of the 2-6-0s' later history. Certainly I have seen it written that American engines of the period were only built to last as long as the boiler. How true this is I can't say, but remembering the opposition the moguls received in this country does make me have reservations.

PRODUCTION

Be all that as it may, however, orders for the locomotives were placed in America and duly completed. The engines were built and erected at their respective factories and tested by raising them on jacks and running them in steam before taking them apart again and shipping them, effectively as kits or parts, to Derby. The contract supervisors in America for the Midland Railway were J.W. Smith, who later became chief locomotive draughtsman under Johnson on 1st January 1901 at Baldwins, and Cecil Paget, later general superintendent, at Schenectady. The first of the Baldwins arrived in a collection of wooden crates via the Manchester ship canal and the Midland Railway system on 24th May 1899. According to Charles Jones, the principal parts of each engine were packed in crates which were marked on the outside with reference numbers, corresponding to those on the consignment notes, so that the contents were readily identifiable, not only as to their description but also the particular locomotive to which they belonged. Twenty-four crates were required for each engine, plus 'twenty-four other cases holding a miscellaneous assortment of fittings to be used on all the engines indiscriminately'. The crates varied considerably in size and weight, the largest being that containing the boiler at 20ft × 10ft × 5ft, and weighing about fourteen tons, whilst the smallest was 5ft × 2ft × 1ft and weighed roughly 3 cwt.

At Derby, waiting to receive the first kits, were engineers previously sent over from the Baldwin factory in Philadelphia. Space in Derby works was at a premium and the only indoor site available was part of No. 8 erecting shop. As the available space in the shop was restricted, it was decided to allocate it for the erection of the Schenectady engines, only ten in number, when they arrived, so the Baldwin engineers had to work out in the open. I don't know what the weather was like in the late spring and summer of 1899, but at least it wasn't winter, and by August all the locomotives had been shipped over. The space cleared for the erection of the Baldwins was described as two sidings in front of the offices, an area traversed by the long footbridge to the works and no doubt overlooked frequently by many of those whose working lives would be affected by the strange new acquisitions. One such individual, driver James Hardy, didn't think much of his first glimpse of them, describing the contents of the crates as displaying 'very rough workmanship'. Such a statement is at odds with contemporary accounts of writers' visits to the Baldwin works, in which they speak of the high standard of workmanship seen, but at this distance in time it is impossible to judge these conflictions. The site was equipped with fitters' benches constructed especially for the task, and with hot water troughs in which the grease and beeswax coating the brightwork (for protection from corrosion in transit) was removed.

Once the crates were unpacked, the cylinders were placed on wooden baulks laid across the rails, the boiler swung into place by steam crane (possibly Cowans Sheldon 15-ton curved jib crane No. 27 of 1893), laid on more timber packing, and the two assemblies bolted

───────────────────────────────

12. This statment was also made by Jones in 1899, but was reportedly something which had been said to him by S.W. Johnson.

An overall view of the Baldwin assembly line outside Derby Works in the summer of 1899. It is interesting to note that the crane being used was hand-operated and not, as referred to in contemporary accounts, a steam type. The piles of components and fittings on the ground were some of the contents of the general crates which contained parts not allocated to specific engines. The fitter in the right foreground was working at one of the benches erected especially for the job.   BRITISH RAILWAYS (DY 1134)

A Baldwin mogul being assembled under the Derby Works footbridge. Note the boiler support plates and running board mounting brackets. The steel bands for retaining the lagging had been placed around the boiler ready to be fitted over the magnesia blocks. The large spring fitting in front of the buffer beam was the mounting for the steam brake operating lever.
COLLECTION R. J. ESSERY

together. The bar frames were then bolted to the cylinders and fixed inside the expansion brackets on either side of the firebox before the drag beam, footplate and steam brake cylinder were fitted. Next came a flurry of simultaneous work fitting the motion, steam and exhaust pipes, smokebox door, chimney, firebox backplate fittings and sanding gear, before the engine was lifted by jacks, more packing placed underneath, and the wheels put in place. Once on its wheels, the engine had the motion connected, coupling and connecting rods put on, and brake gear fitted, before the sheet metalwork of cab and running plate, and the boiler lagging, were applied.

The boiler lagging was described in Cassier's Magazine as 'Magnesia sectional lagging — a white solid substance, exceedingly light in weight, but very tough'. It was supplied in blocks about two feet long and six inches wide, of varying thicknesses to suit the particular part of the boiler it was intended to lag. The barrel was formed from a series of telescopic rings riveted together, widest at the

firebox and narrowing in four steps to the smokebox. Thus, as the external clothing was parallel as far as the smokebox, the lagging got thicker towards the front of the boiler to fill the increasing space. To lag the boiler, wire loops were fastened at intervals along the barrel, the magnesia blocks cut to shape with a saw, and fitted in place, commencing with a line under the centre of the boiler and working up and around. They were attached to the wire loops and to each other by small hooks before being fastened securely in place by steel bands. It was reckoned to take a team of six men 4–5 hours to lag a boiler. Once this was finished, the iron clothing panels, handrails etc. were fitted. In tandem with the erection of the locomotive, the tender was also assembled. The two bogies were placed on the rails, the tender frames lowered on to them, and buffer and drag beams bolted on. The brake gear was fitted, then the tank swung into position by crane and fixed in place before feed pipes, etc, were added. Once engine and tender were complete, they were

A rear view of a part-erected Baldwin 2−6−0. As shown in this photograph, the boiler, as well as having stays from the front platform to the smokebox, was also stayed from the firebox to the rear of the frames. As in the previous photograph, the lagging bands can be seen loosely placed round the boiler. Note the massive steam brake cylinder casting under the cab floor and the direct-acting fore and aft regulator.

BRITISH RAILWAYS (DY 1140)

The sectional magnesia block lagging being applied to the boiler of a Baldwin mogul with three of the fixing bands in place. The process of lagging the boilers is described in the text. The fitters are shown standing on planks laid across the running board suspension brackets. Although only two men were working on the locomotive in this photograph, the normal team employed to lag a Baldwin boiler was six.
BRITISH RAILWAYS
(DY 1147)

A Baldwin nearing completion with cab and external clothing in place. The rear equalising beam can be seen pivoted on the top frame member and, above it, the reversing rod bracketed from the running board.
BRITISH RAILWAYS (DY 1142)

The cab of a Baldwin 2—6—0 under construction, showing the wide rear fairing and the door opening in the cab front. The large difference in height between the running boards and the cab floor is apparent. Note the Midland pattern firedoor.   BRITISH RAILWAYS (DY 1138)

No. 2510 after completion at Derby in 1899. The American origins of the engine are much more obvious than with the Schenectady moguls. The rivet lines on the tender show the outline of the tank, different from the horseshoe type fitted to Great Northern and Great Central Baldwins, and the full depth front bulkhead of the first batch (2501-2510). The toolbox stretched right across the rear of the tender. The balance weights on this batch of locomotives were the same on all the coupled wheels, covering four spokes.
BRITISH RAILWAYS (DY 1151)

Another view of 2510 taken in 1899, included largely because it shows what appear to be polished, unpainted buffer housings. At first glance, it would seem that the shine may have been caused simply by reflected sunlight, but a close study of the original photograph reveals an absence either of lining on the housings or a colour difference between them and the buffer shanks. The baseplates can also be seen to be the same colour as the buffers. One other photograph — that of No. 2506 after completion at Derby — shows the same phenomenon. No. 2510 was the first of the American moguls to be withdrawn, being taken out of service in November 1908. BRITISH RAILWAYS (DY 1149)

coupled together and steam raised before a short trial trip was made and the locomotive put into service.

The new engines certainly looked impressive and some of the doubting Thomases were somewhat mollified by their appearance. Favourable comments particularly were made about the cabs, vastly different from the visually appealing but fairly skimpy Johnson variety. With their long, protective all-over roofs, large glazed side windows, and comfortably upholstered forward facing driver's and fireman's seats, they promised to make the Midland enginemen's lives a little more comfortable, or at least a little less harsh. Later on, though, there were criticisms of these same cabs, as we shall see in due course.

THE MOGULS IN SERVICE
Assembly of the engines was rapid and the first ones entered revenue-earning service in June of 1899. Their principal use was on the Toton to Brent Sidings (Cricklewood) coal trains although they were tried out over most of the Midland Railway system. In July of 1899 Baldwin No. 2503 was operated over the line from Derby to Manchester when it was noted as reaching the top of the climb to Peak Forest with about ninety pounds per square inch boiler pressure remaining. Being fitted with the vacuum brake, and equal in power to the standard Johnson 0-6-0 goods engines, they were, of course, available to be used on a wide variety of trains (They were later placed in power class 2.) The only evidence I

have seen, however, of a mogul on anything other than a goods or mineral train is a single photograph of one of the Schenectady 2-6-0s with what I am told is a train of empty coaching stock at Birmingham New Street after the engine had been fitted with a Deeley smokebox door and Salter safety valves. Driver Hardy had charge of Baldwin No. 2503 for a 1,000 mile trial when the engine was first put into traffic and initially commented that 'The engine looks considerably better now it is in working order' (compared with the 'very rough workmanship' he had noted when the pieces were first unpacked). But by 8th August he had finished the required mileage and was 'very glad to get rid of her'.

The controversy which had greeted the Midland Railway's ordering of the American engines continued after their arrival, encouraged no doubt by comments such as the above, and the detractors must have been delighted when rumours began to circulate of high running costs and maintenance difficulties. Adverse comment and criticism, not just of the engines, but of the Midland Railway's misguided policy (in the critics' eyes) of buying them in the first place, reached such proportions that the Midland's locomotive superintendent, S.W. Johnson, took the unusual step of giving a statement to the press early in 1901. Johnson was reputed to be a normally taciturn man, little given to public statements, but in this instance he was forthcoming on the question of the American engines' performance, maintenance, and costs. The text of his statement is reproduced below and is notable as much for what is implied as for the information it contains:

'To begin with, these American engines are heavier in fuel, oil, and repairs than our own. The orders were given in February 1899 and the engines were delivered in the second half of the same year. Now those engines were not at all English engines built in America. We laid it down that they were to be of the same power as the standard Midland goods engine, and there were a few small details to which the manufacturers had to conform; but generally speaking, the Americans had a free hand, and the engines were for the most part of their own design and pattern, and made in their own way. When they arrived we put them on to our mineral trains running between Toton Sidings, Wellingborough, and London, and set them to do the same work our own standard engines are doing. In January of last year we commenced a six months comparative test, terminating at the end of June, between these Americans and our standard Midland goods engines, built by Messrs Neilson, Reid and Company of Glasgow and Messrs Kitson and Company of Leeds. The two types were set to draw similar mineral trains under the same conditions, and a careful account was kept of the total mileage covered by each, the total coal consumed, and the charges for repair which belonged to each engine. The result was conclusive and is briefly as follows:

Extra working cost of American engine over English (sic) engine.

Fuel	20–25%
Oil	50%
Repairs	60%

'It must be said for the foreign engines that they worked their trains satisfactorily, but their inferiority on the three points named is, on the above showing, incontestable. I never had any doubt in my own mind as to which was the better engine of the two. As to the possibility of repeat orders, I can say nothing beyond referring you to the result of the test.' [In other words, not a chance.] 'Each American engine cost us four hundred pounds less than did those for which contracts were given to British firms immediately in front, and at such a reduction the American engines were put free on our rails just as English engines were. Then there is another point. As I have shown, the Americans were delivered here within a few months of the order being given, yet some contracts which we let out to British firms in 1897 were not completed until February 1900. Of course, this is largely the fault of the engineering strike which caused us to put our work out to America.'

Jubilation in the critics' camp and much comment of the 'Told you so' variety! Now whilst I accept unreservedly the stark results of the trial as reported by Johnson (despite his apparent inability to differentiate between 'English' and 'British'), he was after all a pillar of the church and said to be of unimpeachable integrity, I can't help wondering what the result would have been in the American engines' own country.

CONSTRUCTION
Before going into that, however, it will be worth describing the engines in some detail, starting with the Baldwin variety.

The frames were built up from bars and thus showed much more of the inside 'gubbins' than did the plate frames of British locomotives. Bar frames, like outside cylinders, were an invention of the Stephensons which had been exported to the United States. They were easier to build than plate frames, particularly before iron could be rolled in large enough pieces to enable frames to be cut from a single plate, and had remained as standard American practice after the method had been dropped by British locomotive builders. In the case of the Baldwin engines, the longitudinal bars were of hammered iron 4in deep and 3 or 3½in thick, depending on the particular part of the frame in which they were used. The horns were forged solid with the upper and lower bars and the resultant built-up frames were joined together by a series of transverse bars and plates. At the front end, two single bars were bolted and keyed to the frames, and on to these extensions were fastened the front buffer beam, pony truck pivot and socket, and the cylinders. This system of bolted-on extensions forward of the front hornguides (or 'pedestals' as they were known in American parlance) was designed to allow easy repair in the event of front end damage in an accident as the damaged parts could be removed and replaced relatively easily.

The method of mounting the boiler was quite unlike British practice where the only connection to the frames was normally via the front end assembly and the firebox (in later practice the firebox was mounted on slides so as to take up the boiler expansion and avoid stresses being set up). In the American engines, the boiler was bolted to the frames at three points: the front end assembly; the motion bracket; and an intermediate bracket just ahead of the firebox. The fixings were actually effected via ⅜in thick iron plates between the brackets and the boiler fabric, angled slightly towards the front so that boiler expansion when the engine was in steam would move them to the vertical. The perceived benefit of this arrangement was apparently to distribute the weight of the boiler more evenly over the frames. The smokeboxes were of plain cylindrical form mounted on saddles, and were stayed to the front platforms by heavy round-section bars. Because the frames were thicker than the normal Johnson variety, the firebox was narrower than that on a standard Midland Railway locomotive boiler, only 2ft 9in, and this may have had some bearing on the engines' short life-span, as I shall discuss later. A feature of the fireboxes was the generous water space all round, and they were stayed to the rear frames inside the cab by flat-section bars.

Two sets of Stephenson valve gear between the frames drove the valves via rocking shafts and long valve spindles, criticised by some British observers as being too light for the job. The valves, mounted in chests with easily removable top covers above the outside cylinders, were, however, said to be so well balanced that they offered little resistance to the motion of the spindles and so did not need the latter to be any more robust than they were. The balancing was achieved by springs bearing on the steam chest covers and fitted into cast-iron inserts on the valves, themselves cast iron, to equalise the steam pressure on the valves and so make them free to slide. Lubrication of cylinders and valves was from the normal arrangement of sight feed lubricators.

Lubrication of slidebars, connecting rod bearings, coupling rod bearings, and eccentrics was quite different from the usual Midland Railway practice of having a piece of worsted dipping into an oil cup and transferring oil through capillary action to a tube leading to the slide or shaft. By withdrawing the worsted trimming, the flow of oil could be cut off when the locomotive was out of use. The American method was to lead the oil down through a tube, into which was mounted a tapered rod which could be raised or lowered by a set screw, thus adjusting the size of the opening and so regulating the flow of oil. The driver was expected to mark the setting of the screw which gave the degree of lubrication required, with the recommendation that summer and winter settings be so marked. Ways of cutting off the flow of oil when the engine was out of service varied according to the location of the lubricator. In the case of the connecting, coupling, and eccentric rods, the feed tube came halfway up the inside of the reservoir, oil was put in up to this level, and the motion of the rod or eccentric splashed oil over the top of the tube. Thus when the engine was stationary, there was no oil getting to the bearing surface. The slide bars and piston rod lubricators, however, relied on the driver screwing the set screw fully down to cut off the oil supply and so prevent waste.

The Bissell truck frames were built up from bars with a central casting projecting up into a socket which protruded above the front platform of the locomotive. The swing links and axlebox springs were as described previously. The wheels were 8-spoked and 2ft 9in in diameter. The frames were extended rearwards in a triangular structure, the apex of which was pivoted on a cast-iron pin just ahead of the leading coupled wheels, and the weight on the truck was equalised with these wheels. The equalising beam bore at the front on the centre truck casting, and at the rear on a transverse member between the front spring links of the leading coupled wheels. The pivot point of the beam was attached via a pin to a bracket mounted between the cylinders. The driving wheels and rear coupled wheels were also equalised by beams on either side outside the frames, fastened to the rear spring links of the drivers and the front ones of the rear wheels. The pivots were pinned to mountings set into the upper frame longitudinal members.

The axles of the 5ft diameter 15-spoked coupled wheels were carried in bearings of quite generous proportions, by Midland Railway standards, the journals being lubricated by pads on the underside dipping into reservoirs beneath them; the axleboxes were supported by leaf springs mounted above the axles.

The boiler was surmounted by three dome-like structures, the middle one of which actually served as the dome cover and mounted a pair of uncased Coale safety valves set to 160 pounds pressure. The rearmost casing contained another spring balance safety valve and mounted the whistle, described as a typical American hooter rather than the more familiar eldritch shriek of the Johnson variety. The forward mounting was a large sand box, positioned above the boiler in order to keep the sand warm, dry, and free flowing. In American motive power depots these sand boxes were filled from a gantry over the track, but the Midland Railway had no such arrangements at its running sheds, and in service the boxes proved awkward to fill, although the rear sandboxes were mounted below the running plate in normal British fashion and did not present a problem. Boiler-mounted sandboxes had been deleted from British locomotive practice because of the tendency for sand spilled when filling them to get into the motion and so cause wear on the moving parts. The Great Northern and Great Central Railway Baldwins had all the sandboxes mounted below the running plate and so avoided any such difficulties.

A photograph of the first Baldwin mogul delivered to the Midland Railway, reproduced on page 18, shows the engine with a polished copper cap to the chimney, and it has been supposed by some people that all of them were so equipped. David Tee assures me that he has seen contemporary correspondence referring to such chimneys having been seen on Midland Railway Baldwins. The photographic evidence, however, does not seem to support this. Pictures of the engines being erected at Derby Works in 1899 do not, at least to me, show evidence of copper caps (unless they were painted — in my opinion an unlikely complication) and I have never been convinced by assertions that some photographs of moguls in service show such chimneys. It is quite possible, of course, that the first locomotive built for the Midland by Baldwins had a copper-capped chimney, either for publicity purposes or as a standard fitting, and that subsequent engines were not so equipped. It is also possible that I am entirely wrong and that all the Baldwins were copper tops! Unfortunately, I do not have sufficient evidence to make a definitive statement except to say that it would appear that at least one of the moguls ran in this country with a copper cap to its chimney.

The cab, as mentioned before, was much bigger than anything previously seen on the Midland Railway, with upholstered seats for the crew and an extensive roof designed to keep the fireman dry in bad weather. The large side windows were glazed and could be slid open. In addition there were windowed doors in the cab front, where the spectacle plate would normally be on a British locomotive, so that the driver could get out onto the running boards to inspect or oil around the engine whilst it was in motion without any gymnastics around the cab side sheets. In fact, on the Baldwin moguls this would have been a very tricky manoeuvre due to the difference in height between the cab floor and the high running boards. Two other windows in the cab front could also be slid open. The rear of the cab was partially enclosed by a fairing which stretched across under the overhanging roof and down the sides.

It would be imagined, therefore, that the Baldwin cabs would be popular with Midland enginemen, and at first glance they seemed to be. Unfortunately, the cabs also contained about 2ft of firebox and so they got very hot, probably no bad thing in the middle of Mallerstang Common in winter, but they must have been uncomfortable in warm weather. Many of the cab fittings were standard Midland Railway items, such as the firedoor and steam brake, but the regulator was a horizontal lever operated in a fore and aft sense and moved a spindle through a gland in the backplate, which then acted directly on the steam valve. The release handle on the reversing lever was the opposite way round from the normal British practice, being to the rear of the lever.

The tender was mounted on two bogies with 3ft diameter 10-spoked wheels. Tank capacity was somewhat greater than the usual Midland Railway goods tender at 4,000 gallons. Apart from the fact that they were fitted with Johnson pattern coal rails and standard Midland couplings and buffers, they were almost standard Baldwin products. Unlike the Great Northern Railway and Great Central Railway's Baldwin tenders, however, the Midland examples did not have horseshoe tanks, as evidenced by the rivet patterns on the tank sides. This necessitated a relocation of the brake gear pull rod to the outside of the leading bogie on the left-hand side.

There were some variations between the first ten Baldwin engines to be delivered and the later batches. On the earlier locomotives, the balance weights on the driving wheels were all the same size, each covering four spokes opposite the crank pin, whereas the subsequent batches had larger balance weights on the middle driving

The first stage of erection of a Schenectady 2–6–0 in No. 8 shop in Derby Works during the summer of 1899. The boiler had been attached to the cylinders by the two-piece steam chest casting and smokebox saddle which were bolted together along the centreline. The brackets at the rear of the steam chest for attaching the frames can be seen lying on the floor beyond the boiler.
BRITISH RAILWAYS (DY 1158)

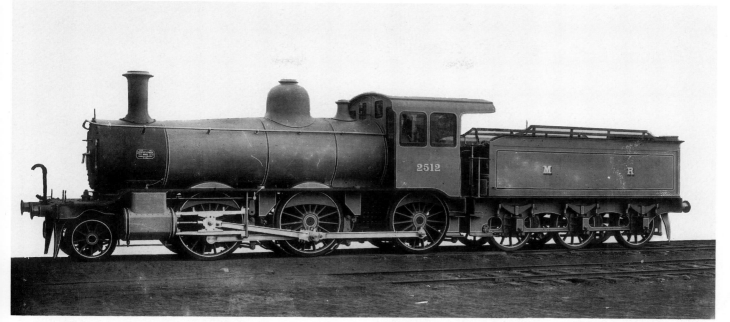

A Schenectady mogul photographed in works grey livery at the factory in America, probably early in 1899. The smokebox front differs from that seen when the engine was delivered to the Midland, having a ring of prominent bolt heads around the periphery. The valances curving out from behind the cylinders to near the ends of the buffer plank can be discerned by close inspection. The locomotive was Schenectady Works No. 5038, Midland Railway No. 2512 until 1907 when it became No. 2231. Note the very light rails, on which the locomotive was standing, spiked directly to the sleepers.

BRITISH RAILWAYS (DY 1162)

wheels which encompassed six spokes (in Midland Railway parlance all coupled wheels were referred to as driving wheels, not just those onto which the connecting rods drove). This feature was in common with the Great Central Railway's Baldwin engines; the Great Northern examples were different again and had 5-spoke balance weights on all axles. Why these variations should occur in locomotives which appear to have been designed to be the same dynamically I have no idea. The position of the dome-mounted safety valves also differed; the first ten engines delivered had them mounted to the rear of the casing whilst on the rest they were moved forward. The earlier tenders had a forward bulkhead which extended the full depth of the tank sides and a large transverse toolbox across the rear, whereas the bulkheads on later ones were shorter and toolboxes were fitted longitudinally on the right-hand side at the front. All except the tenders coupled to the first ten locomotives had an angle beading strip along the bottom of the tank sides connecting to the platform. There were variations in the riveting of the coal rail supports which do not seem to have been consistent on any particular batch of tenders.

The Baldwins looked what they were, typical American 2-6-0s, but the Schenectady moguls were semi-Anglicised, a fact which attracted much comment when they were first seen. More recently Professor Tuplin wrote that he 'hazards the opinion that they may well have been the most elegant-looking locomotives ever to have come out of America'[13], an opinion with which I wholeheartedly concur.

Like the Baldwins, they were shipped over as kits of parts and erected at Derby by American engineers. In construction they were very similar to the Baldwins but there were notable differences. The frames, whilst basically of the same bar construction as their brethren from Philadelphia, had deeper side extensions forward of the leading coupled wheels and valances which curved outwards from behind the cylinders to just inside the extremities of the buffer planks. Whereas the Baldwins had high running boards and separate front platforms at a lower level, the Schenectadies were fitted with wholly more elegant curved running plates sweeping from front buffer beam to the rear of the cab. The boiler was tapered forward of the middle ring to a smooth join with the smokebox rather than being stepped like the Baldwins and there were no stays from the smokebox to the front platform. Cylinder cladding was reduced on the outsides of the locomotive to clear the Midland loading gauge, giving the cylinders a flat-sided appearance. The sandboxes were positioned below the running plate in normal British fashion, doing away with the inconvenience of filling the boiler-mounted type and the problems caused by stray sand. Although the Bissell truck wheels were slightly larger than those of the Baldwins, they were 3ft in diameter and had nine spokes; the coupled wheels were the same. The locomotive springs were equalised like those on the Baldwins, but the equalising beams were pivoted on brackets fixed to the lower frame members and the springs were below the axles.

The dome casing was a massive affair mounted on the middle ring and contained a pair of pop safety valves, but on the Schenectady moguls it didn't look out of place. A further pop valve was positioned on the rear ring just ahead of a Johnson type whistle. The cab was very similar to the Baldwin version with front doors and sliding glazed front and side windows, although the rear fairing

---

13. Professor W.A. Tuplin wrote widely on steam locomotive engineering, practice and history. The statement quoted appeared in *Midland Steam* published in 1973.

In this picture the frames, front frame extensions, and motion bracket had been added to the boiler assembly along with some pipework. Also evident from close inspection are the wire loops to which the boiler lagging would be fixed. The builder's plates were welded to the smokebox.
BRITISH RAILWAYS (DY 1155)

Further progress is evident in this photograph. As can be seen, the firedoor was a standard Midland type and the rear running plate was bracketed directly to the firebox. The firebox was mounted on sliding brackets resting on top of the bar frames and the equalising beam pivots were bracketed to the lower frame member. The valve spindle rocking lever can be seen just behind the motion bracket and the twin pop safety valves were mounted on the dome. Note that the reversing lever release handle was to the rear of the lever unlike that on other Midland locomotives with lever reverse.
BRITISH RAILWAYS (DY 1157)

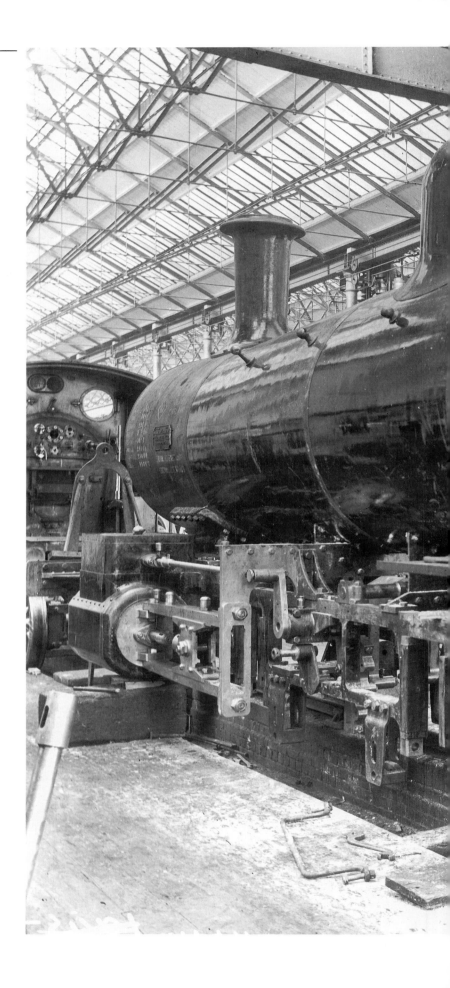

This view shows the addition of boiler cladding, boiler fittings, firebox backplate fittings, cab splashers and part of the platforms. The brake hanger brackets can be seen bolted below the lower frame members. Note also the intermediate boiler support plate between the firebox and the motion bracket.
BRITISH RAILWAYS (DY 1154)

Another Schenactady mogul being erected at Derby in a fairly advanced stage of completion. The pony truck pivot can be seen projecting through the front platform just in front of the smokebox. Note also the brackets supporting the buffer beam from the forward frame extensions and details of the smokebox saddle. As is evident from the photograph, the engines were painted before being shipped from the United States, witness the lining and lettering on the buffer beam, and only needed touching up after reassembly.

BRITISH RAILWAYS (DY 1156)

was not as wide or deep, and had the same advantages and drawbacks; the amount of firebox intruding into the cab was even more than the Baldwins and amounted to nearly 2½ft. The Schenectadies made much more extensive use of flush rivets than 'did the Philadelphian locomotives, which were liberally sprinkled with snap-head rivets, particularly around the cabs and tenders, and thus looked much cleaner. They were nearly four tons heavier than the Baldwins, tipping the scale when in working order at just under 50 tons.

The tender was almost a standard Johnson 3,250 gallon type, although this wasn't specified as a requirement when the contract was awarded, the only real differences being slight alterations in wheelbase and tank width, and the absence of vertical beading halfway along the tank sides. Despite looking smaller than the Baldwin bogie tenders, the Schenectady 'Johnsons' were 2½ tons heavier.

Since they were erected inside Derby Works, the Schenectady engines did not require the hastily contrived benches, nor use of steam cranes, that characterised the Baldwin production line.

All in all, the Schenectady moguls were not that different in appearance from locomotives which could be seen, either at the time of their arrival or shortly afterwards, running on other British railways and built by British manufacturers. Both types of moguls received the full Johnson lined red livery and, in my view, it suited them very well, particularly the New York engines.

MODIFICATIONS

By the late autumn of 1899, all the American engines were in service. The first ten Baldwins were numbered 2501 to 2510 and were sent after trials to Toton. The Schenectadies came next in the numbering sequence, occupying the block 2511 to 2520, and were all allocated

A Schenactady Bissell truck showing details of construction and springing. Construction and operation of a Bissell truck is described in the text. The large pin in the central casting, which protruded into the socket in the front platform, is evident. BRITISH RAILWAYS (DY 1161)

The rear of a completed Schenectady mogul. The cab rear fairings were not as extensive as those of the Baldwins, but still proved unpopular with enginemen. Note the upholstered seats for the crew.
BRITISH RAILWAYS (DY 1159)

to Wellingborough shed. Finally, the last two batches of Baldwins became Midland Railway numbers 2521 to 2540 and were equally divided between Leeds and Sheffield.

Modifications to the American engines during their service with the Midland Railway were few. At least one of the Baldwins was fitted with a flowerpot type chimney sometime between 1902 and 1905, and two at least had modified buffer beams, the round-ended Baldwin ones being replaced by single plate rectangular versions. Some of the Schenectadies acquired Johnson's elegant cast iron chimneys in the same period. In neither instance, however, can I say how many locomotives were so modified. All I can do is cite specific cases seen in photographs. Details of such engines are contained in the captions to the plates. Salter safety valves appeared behind the domes of at least two of the Schenectadies, but the only photographs I have seen showing this modi-

fication were taken when the locomotives were in Deeley livery, so I don't know whether he or Johnson ordered the alteration, nor whether any others were so treated in addition to the ones mentioned. The moguls received Deeley style smokebox doors at some stage, and Deeley parallel chimneys with capuchons appeared on them after 1907; neither (in my opinion) did anything for their appearance.

Liveries were mixed. As previously noted, when new they were turned out in lined red, but photographic evidence suggests that some of them wore unlined red or, possibly, locomotive brown at various times, and later on some may even have been black. They were not treated as a class, however, as some later appeared in the simplified Deeley lined red livery whereas others were plain at around the same time. The lining styles also varied, some engines having lining around the cylinders and

A Midland Railway official photograph, later used in postcards and publicity material, signed by S. W. Johnson, probably as an approval for its use. The large amount of firebox included in the cab is apparent, as is the extent of the cab roof which would have protected the fireman when shovelling coal. The difference in tender wheelbase from the normal Johnson 3,250 gallon type, and the absence of a central vertical beading strip on the tank side, are also evident.
COLLECTION
R. J. ESSERY

The same photograph as that shown above, this time after artistic retouching for a Midland Railway postcard. This plate is included to show how apparently official material can be misleading. The retoucher had altered the profile of the valancing forward of the cylinders, and made the cylinders themselves appear to be more slab-sided than they actually were, with square lower corners to the cladding. The cab side windows have been given sharp lower corners and the dome has a distinct taper. These errors have been incorporated into more than one illustration of a Schenectady mogul over the years. The rather improbable sky background added by the artist makes the engine appear to be in a stage set.
AUTHOR'S COLLECTION

Another of the first batch of Baldwin moguls, No. 2505, photographed at Bedford. From this view there is no doubt that the buffer housings were painted and lined, although, unlike No. 2510, the cylinders and valve chests are unlined.   **COLLECTION J. BRAITHWAITE**

valve chests whereas some did not, even, apparently, from new. Unfortunately, not all the engines received attention from photographers and a relatively small number, particularly of the later batch of Baldwins, appear in a lot of pictures whilst some are absent from the photographic record.

OPERATION AND MAINTENANCE
But the fact remains that they were, as we have seen from Johnson's statement, expensive to operate and maintain, and did not enjoy what was, by Midland Railway standards, a reasonable life span — the longest lived of them lasted a mere fifteen years or so. In attempting to explain why this should have been, I do not pretend to have all the answers, nor to defend them unduly, but I think there may have been some mitigating circumstances.

As I have tried to show, they were not popular among the railway cognoscenti nor the establishment, and at least some of this feeling must have been prevalent among Midland Railway operating and maintenance staff. I have to wonder, therefore, whether they received a fair chance. Added to this they were unfamiliar in many respects to the Midland enginemen and fitters, and I would hazard a guess that this unfamiliarity could in itself have led to problems. Men used to the standard Midland Railway locomotive types would not, I suggest, have necessarily got the best out of these strange and controversial machines. There are sufficient instances of other new and slightly revolutionary or unusual locomotive types having teething troubles, due to differences needed in handling and maintenance techniques, to suggest to me that the moguls may well have suffered in this respect. When it is remembered that the bad press they received wanted, and indeed expected, them to fail, it can be recognised that they very probably did not have the best start in life, and their comparative trial, as described by Johnson, occurred only a year after they entered service,

The fireboxes, as previously described, were narrower and deeper than was normal on Midland Railway standard locomotives. It is quite probable that if they were not fired carefully, a thick fire would result and coal consumption would go up, and the firing technique required to get the best fuel consumption from them may not have been natural to Midland firemen. There was an extra axle to lubricate and an extra set of sliding surfaces in the rocker actuated valve gear, so that could explain some of the increased oil consumption. Because of the (to Midland men) strange method of lubrication, it is possible that drivers, used simply to putting in and taking out worsted trimmings at the beginning and end of a duty, would have experienced difficulty in getting the screw adjustments just right; and when in doubt it is normally better to have too much lubrication than too little, so it is quite likely that at least some oil was wasted when the engines were running. It must also have been tempting not to mess about with the settings of the adjusting screws on the slidebar and piston rod lubricators at the end of a turn, but just to leave them set for running and accept the consequent oil loss. I do not say that this is an explanation for all the extra consumption of coal and oil noted in Johnson's statement, but merely to suggest that maybe the American engines were not intrinsically as bad as they appeared from the bald facts surrounding their use in this country.

The greatly increased maintenance costs reported by Johnson, however, are more difficult to try to explain. Unfamiliarity and the fact that most of the parts were not standard Midland Railway components may have had some bearing, but there are sufficient comments of rough workmanship to make it unwise merely to reject them as xenophobic. Unfortunately, there is insufficient detail available to make an objective assessment of which areas gave most trouble and resulted in the most markedly increased maintenance costs.

The reaction to the adverse reports in the British railway press resulted in something of a backlash on the other side of the Atlantic, which was echoed to some degree in this country by a few impartial observers.

No. 2505 again (this engine seems to have been one of the photographer's favourites) seen this time with both cab front doors, the sliding side windows, and the cab roof ventilator open. As previously noted, the cabs of the American locomotives could get hot. The date of the photograph is unknown but is some time between 1902 and 1907, and the painting and lining scheme was unchanged. The long, slender valve spindle is evident in this view and it is understandable that to some people they may have looked too thin. COLLECTION R. J. ESSERY

The photographer's friend, No. 2505, taken prior to 1902. As can be seen, even the guard irons were bolted on and were thus easily replaceable in the event of damage. The polished valve chest and cylinder covers also stand out and it is obvious that the axlebox lids on the tender were unpainted, as were the smokebox door hinge and, apparently, the smokebox stays. COLLECTION R. J. ESSERY

No. 2505 on an up freight at Welsh Harp between 1899 and 1902. The cluttered appearance of these engines when compared with contemporary British types is evident in this view.
COLLECTION R. J. ESSERY

Charles Rous-Marten pointed out that in America the 500-ton trains on which the moguls were often rostered by the Midland would have utilised bigger motive power; the Midland Railway 2-6-os were, by American standards, small locomotives. In other words, the Midland worked them harder than would have been the case were they employed in the United States. Although this is no doubt true (I am not sufficiently well informed on turn-of-the-century American railway operating practice to judge), it still has to be said that it was specified to the manufacturers at the time of ordering as to the power requirements the Midland Railway would expect of them. It is ironic that the Midland, often criticised by its rivals for not utilising locomotives to their maximum

No. 2515 in the summer of 1900 showing the driver's side cab door open and the right-hand cab front window slid aside. The amount of firebox within the cab could make the footplate very hot and the doors must have been a welcome feature, apart from the easy access to the running plate they provided.
AUTHOR'S COLLECTION

Same engine, same day as the photo above, now attached to a coal train waiting to head south. The exposed oil pipe along the side of the valve chest was an early modification to some engines, although I cannot find any record of when it was carried out, nor whether all locomotives were so modified. The elegant tapered chimney and massive dome casing are shown to good effect in this picture. COLLECTION J. BRAITHWAITE

potential, should be accused of flogging engines. Some commentators suggested that the quality of materials used in the engines' construction was inferior to that employed by British manufacturers, but without a proper qualitative assessment it is impossible to judge. Rous-Marten said that the philosophy of American railway companies was to buy locomotives cheaply and run them until they wore them out without bothering to nurture them the way British companies did.[14] If this was indeed the case, I still can't see that it would be a valid reason

for using inferior materials; in fact, I would have thought quite the opposite. Most of the comment, critical or otherwise, on the American engines seems to have been made about the Baldwin machines. Whether this was

14. Charles Rous-Marten was a New Zealander who was familiar with both British and American railway practice. His remarks on the Midland Railway's American engines were made in *Railway and Locomotive Engineering* in 1901 in response to the backlash caused in America by the adverse British comment on their performance.

No. 2511 on a southbound mineral train in 1900, a duty on which it probably spent most of its working life. Like its brethren from New York, this engine was based at Wellingborough, but was the first Schenectady to be withdrawn, by then numbered 2530, in August 1911.

AUTHOR'S COLLECTION

The left-hand side of No. 2517, still without external oil pipes to the valve chest.

COLLECTION R. J. ESSERY

No. 2521 was the first example of the second batch of Baldwin moguls to be delivered, and the longest-lived of the type, lasting until May 1914. Comparison with the photo on page 26 will illustrate the differences between the batches. The middle coupled wheels had balance weights covering six spokes, and the pop valves on the dome casing were further forward than those on the first ten engines. The front bulkheads on the tenders were shorter than the earlier examples and there was a forward mounted toolbox. The brake operating pull rod, relocated because of the change from horseshoe tanks, can be seen running outside the front bogie just below the tender frame. Less obvious is the beading strip along the join between the tank side and the platform and the different rivet pattern on the coal rail supports on tenders fitted to locomotive numbers 2521 et. seq.

AUTHOR'S COLLECTION

No. 2526 photographed at Lancaster between 1902 and 1907. There is no lining evident on the engine and I would suggest that it was finished in plain red livery. The chimney cap appears to have been unpainted and this is suggested by some commentators to be evidence that the Baldwin moguls had copper-capped chimneys. They were wide locomotives, 8ft 4in over the cylinder cladding, and the right-hand cylinder seems to have been in collision with something on the lineside. The rear safety valve was shielded by a semi-circular shroud, a feature I have only seen in photographs of this engine and Nos. 2521 and 2527. The lower tender bulkhead of the second batch can be seen.
COLLECTION S. N. ROSS

because they were the first ones to arrive at Derby I can't say, but certainly the Schenectadies were less well known, and most contemporary, and indeed later, writings about the moguls just seem to lump them into one class.

The final point I would make is that standard Midland boilers would not fit between American standard gauge bar frames. Thus when the boilers were due for renewal, they either had to be bought from the United States, specially made in Britain, or the locomotives scrapped. With a relatively small class of engines, known to be more expensive to operate, and with the added expense of the options to renew the boilers, it would have been difficult to justify keeping them. They were scrapped. The Great Northern and Great Central Railways took the same decision.

THE END

The first mogul to be withdrawn was Baldwin No. 2209, ex No. 2510, which went in November 1908. By the end of 1913, there were only six of the moguls left, three Baldwins and three Schenectadies, and in August 1915

No. 2526 of the second batch of Baldwins, seen some time between 1902 and 1907. There does not appear to be any lining on the locomotive anywhere and it was probably in plain red or even black livery. Even so, the smoke-box stays, cylinder covers, valve chest covers and tender axlebox lids were still unpainted. The safety valve on the rear ring had received a semi-circular shroud fitted in front of it, but the fitting date and purpose of these additions is unknown. Note the jack on the platform in front of the fireman's cab door.
COLLECTION
R. J. ESSERY

An interesting photograph of No. 2534 taken sometime between 1902 and 1907. By this time the engine had been fitted with a flowerpot chimney.
COLLECTION R. J. ESSERY

the last survivor, Schenectady No. 2233 in the 1907 renumbering, originally No. 2514, was withdrawn from Wellingborough. Proportionally the Schenectadies lasted slightly longer; whether they were better machines or not I don't know, but, since I have something of a soft spot for them, I would like to think so. One trace of the moguls lingered on. The Schenectady tenders were stored at Derby and reappeared, slightly modified, in 1917 coupled to the first ten of the 3837 class 0-6-0 'big goods' engines.

It should be remembered, though, that the acquisition of the American engines in the locomotive starved days of 1899 got the Midland Railway, at least partially, out of a hole and provided much needed additional motive power when it was unavailable from normal sources. Bear in mind, too, that there was never any suggestion that when in traffic they did not perform the day-to-day

tasks the Midland set them. As Samuel Johnson said of them, 'They worked their trains satisfactorily', and this must be their epitaph.

The works numbers, original Midland Railway numbers, post 1907 numbers, and withdrawal dates for the Baldwin engines were as shown in Table 4. Those for the Schenectady locomotives are given in Table 5. Leading particulars for each type are given in Tables 6 and 7.

I have to record my thanks to all those people who have helped me over the years to gather information on the two categories of American locomotives used by the Midland Railway, and to understand (I hope) the material I had. In no particular order they are: The late Laurie Ward, Peter Truman, Jack Braithwaite, Steve Ross, Pete Kibble, Tony Wall, Bob Essery, the late John Horton, David Tee, John Edgington, Laurie Knighton, Tim Watson, and the staffs of the Public Record Office,

The Baldwin moguls seem to have lost their attraction for photographers after a while, and this picture of No. 2207 (ex-No. 2508) is one of the few I have seen of them with Deeley capuchoned chimneys.
COLLECTION
R. J. ESSERY

One of the few photographs of a Baldwin mogul with Deeley smokebox door and chimney. No. 2208, ex-No. 2509, is also unusually seen in charge of a cattle train at an unknown location. As frequently was the case, the driver's cab door was open. A close study of this picture suggests that the safety valves on the dome may have been Salter types instead of the Coale valves originally fitted to the Baldwins, but the definition of the photograph and the angle from which it is taken, make it impossible to be certain. Although in Deeley condition, No. 2208 still had lining on the platform valances.
COLLECTION
R. J. ESSERY

The Science Museum, Gloucester Libraries, the British Library, the National Railway Museum, the University of Texas, and the Smithsonian Institute.

Thank you Ladies and Gentlemen. If I have missed anyone out, it is a sin of failed memory and I can only apologise.

In preparing this work I am also indebted to David Ross for producing the drawing of the Baldwin mogul, from which the diagram I have included was prepared, and for pointing out some of the variations in the different batches of engines, even if I still don't agree with him about Baldwin chimneys. Additionally I have made use of the writings of previous authors on the subject although I have tried where possible to verify their statements.

No. 2232, previously 2513, with its pony truck removed in 1909. The locomotive had been 'Deeley-ised' with a rather ugly parallel-sided capuchoned chimney and six-dogged smokebox door with front numberplate. The cabside number had been replaced by the second style of Midland Railway locomotive crest and the lining simplified, although the engine was still in red livery. The original pop safety valves in the dome had been replaced by a pair of Salter valves, a modification not often apparent in photographs of these engines.   COLLECTION R. J. ESSERY

Although not of particularly good quality, this picture has been included as it is the only other one I have seen which shows definitely the modification of a Schenectady mogul with Salter safety valves. It also illustrates, in my view, the unfortunate appearance of the Deeley chimney on these engines. The subject of the photograph is No. 2233, originally 2514, at Geddington on 18th October 1911.
COLLECTION J. BRAITHWAITE

An unknown Schenectady 2—6—0 hauling a long goods train. As Samuel Johnson said, "They worked their trains satisfactorily".

This photograph is interesting for several reasons. It shows No. 2236, seen in two previous plates in its original guise as No. 2517, on the traverser at Derby works sometime after 1907, after having been fitted with a Johnson one-piece chimney, in my opinion a modification which suited it. It had also received the external valve chest oil pipes, and the power classification '2' had been added below the number.

COLLECTION R. J. ESSERY

TABLE 4					TABLE 5			
BALDWIN LOCOMOTIVES					**SCHENECTADY LOCOMOTIVES**			
Works No.	1899 No.	1907 No.	Withdrawal		Works No.	1899 No.	1907 No.	Withdrawal
16622	2501	2200	8/1911		5037	2511	2230	8/1911
16623	2502	2201	9/1913		5038	2512	2231	6/1913
16624	2503	2202	10/1911		5039	2513	2232	4/1913
16625	2504	2293	2/1910		5040	2514	2233	8/1915
16626	2505	2204	2/1910		5041	2515	2234	4/1913
16627	2506	2205	5/1913		5042	2516	2235	7/1914
16628	2507	2206	3/1913		5043	2517	2236	12/1913
16629	2508	2207	11/1910		5044	2518	2237	7/1914
16630	2509	2208	10/1911		5045	2519	2238	6/1912
16631	2510	2209	11/1908		5046	2520	2239	3/1913
16844	2521	2210	5/1914					
16845	2522	2211	8/1913					
16846	2523	2212	3/1909					
16847	2524	2213	4/1913					
16848	2525	2214	4/1914					
16849	2526	2215	7/1909					
16850	2527	2216	3/1913					
16851	2528	2217	3/1913					
16852	2529	2218	10/1913					
16853	2530	2219	3/1909					
16960	2531	2220	4/1909					
16961	2532	2221	4/1909					
16962	2533	2222	3/1913					
16963	2534	2223	3/1914					
16964	2535	2224	5/1913					
16983	2536	2225	4/1909					
16984	2537	2226	4/1909					
16985	2538	2227	7/1912					
16986	2539	2228	4/1912					
16987	2540	2229	7/1909					

Another view of No. 2233, this time taken at Birmingham New Street by W. L. Good in May 1914. As can be seen, the addition of Salter safety valves is not always apparent from photographs. Reportedly, No. 2233 was at the head of a train of empty coaching stock, but why a Wellingborough-based engine should have been on such a duty in Birmingham is not recorded. It is, however, the only photograph I know of that shows an American mogul on a train of passenger vehicles. By the time this picture was taken, there were only two other moguls left, both Schenectadies, and No. 2233 eventually became the last of its line, being withdrawn in August 1915 and so ending an era on the Midland Railway.

AUTHOR'S COLLECTION

TABLE 6
LEADING PARTICULARS — BALDWIN ENGINES

Boiler pressure	160 pounds per square inch
Cylinders	2 at 18 inches × 24 inches
Nominal tractive effort	18,000 pounds approximately
Driving wheel diameter	5 feet
Coupled wheelbase	6 feet 3 inches × 8 feet 6 inches
Pony wheel diameter	2 feet 9 inches
Pony wheel position	7 feet 5 inches ahead of leading coupled wheels
Length overall (locomotive +·tender)	51 feet 1 inch over buffers
Weight in working order	45 tons 16 hundredweight 1 quarter
Tender capacity	4,000 gallons
Tender wheel diameter	3 feet
Tender wheelbase	14 feet 4 inches
Tender bogie wheelbase	4 feet 8 inches
Weight in working order	34 tons 12 hundredweight 2 quarters

TABLE 7
LEADING PARTICULARS — SCHENECTADY ENGINES

Boiler pressure	160 pounds per square inch
Cylinders	2 at 18 inches × 24 inches
Nominal tractive effort	18,000 pounds approximately
Coupled wheel diameter	5 feet
Coupled wheelbase	7 feet × 8 feet 6 inches
Pony wheel diameter	3 feet
Pony wheel position	6 feet 6 inches ahead of leading coupled wheels
Length overall (locomotive + tender)	51 feet 11¼ inches
Weight in working order	49 tons 15 hundredweight 2 quarters
Tender capacity	3,250 gallons
Tender wheel diameter	4 feet 2½ inches
Tender wheelbase	12 feet 3 inches equally divided
Weight in working order	37 tons

MIDLAND RECORD

Midland Record is an exciting part work which builds into an unrivalled record covering diverse aspects of the Midland Railway from its inception, through grouping and into nationalisation. Each 80-page issue is sewn and bound with a laminated card cover. From time to time we will also produce additional supplements covering associated subjects (e.g. engine monographs or line histories). These will be presented in the same format as *Midland Record*, but the number of pages will vary according to requirements.

Midland Record is available through bookshops and specialist retailers or direct from the publishers, but NOT through the newstrade. Copies supplied direct are accompanied by an order form for the following issue. For the time being at least, there will be no regular interval between the publication of each part, so the approximate date of publication of the following issue is announced in each one.

80 pages. ISSN 1357-6399. Laminated card cover.   Price £8.95
Subscriptions: 3 issues for £26.50.

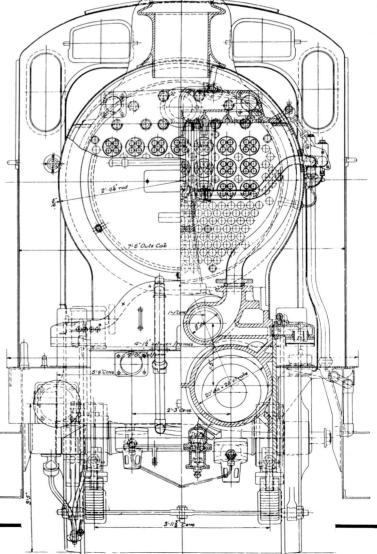

STILL AVAILABLE
Nos. 1-7.   £8.95 each

Please add £1.00 P & P for orders under £10, or £1.50 for orders over £10.

WILD SWAN PUBLICATIONS LTD.
1-3 Hagbourne Road, Didcot, Oxon,
OX11 8DP. Tel. 01235 816478.

M.R. BALDWIN 2–6–0

A Baldwin mogul of the first batch (Nos. 2501-2510) as originally built. Note the equally-sized balance weights and safety valves mounted to the rear of the dome casing which identify the earlier engines. The tender has a large tool box at the rear, a full depth front bulkhead, and no beading along the bottom edges of the tank sides. The drawing from which the diagram was produced was prepared by David Ross from a general arrangement drawing supplied by the University of Texas, who hold most of the Burnham, Williams & Co. archive material. As with many other subjects, however, what was drawn in a G.A. and what was actually built differed in minor details, and extensive reference to photographs was made during the preparation of the diagram. Why the archives of a Philadelphia company should end up in Texas I have no idea.

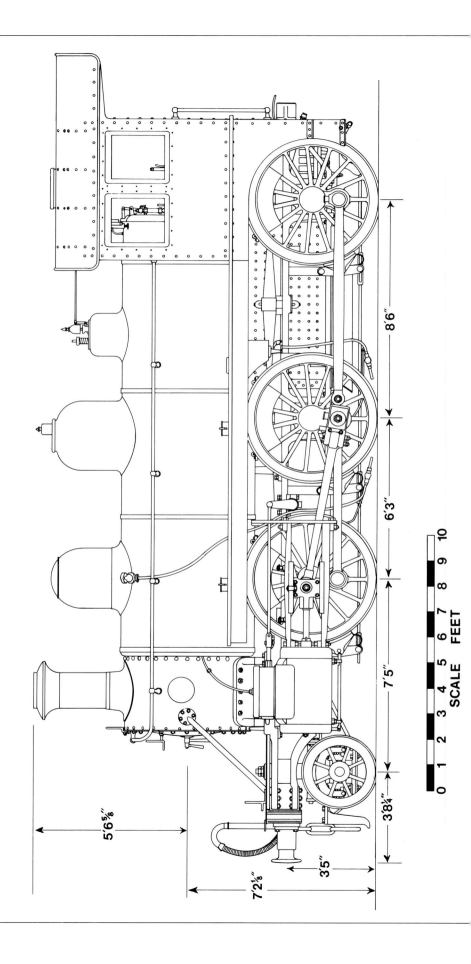

SCALE FEET

0 1 2 3 4 5 6 7 8 9 10

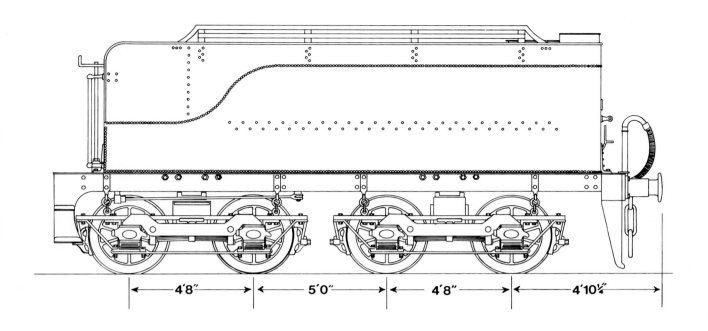

4′8″ · 5′0″ · 4′8″ · 4′10½″

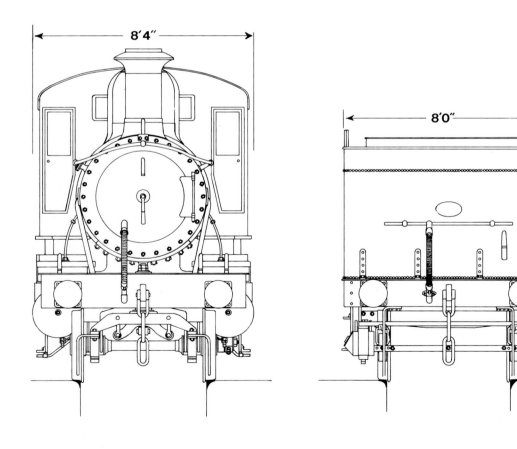

8′4″

8′0″

M.R. SCHENECTADY 2–6–0

A Schenectady mogul of 1899 in its original condition. The main source of reference used in the production of this diagram was a copy of a very battered drawing loaned to me many years ago by the late Laurie Ward. The drawing did not appear to be of Midland Railway origin and was, in Laurie Ward's opinion, produced by the Schenectady Works to show dimensions and major details for the benefit of the Midland, possibly before the engines were contracted for, although there was no date discernible. This view was supported by the fact that there were minor differences between the drawing and the locomotives as they appeared from Derby. Some details, therefore, have been estimated from photographs. There is in existence a Midland Railway weight diagram of these engines, but apart from weights and major dimensions, which agree with the drawing on which I based the diagram, it is hopelessly inaccurate.

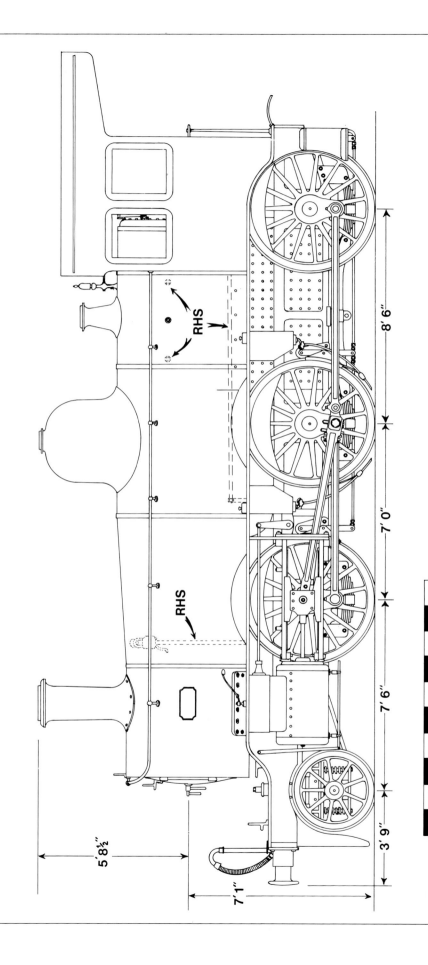

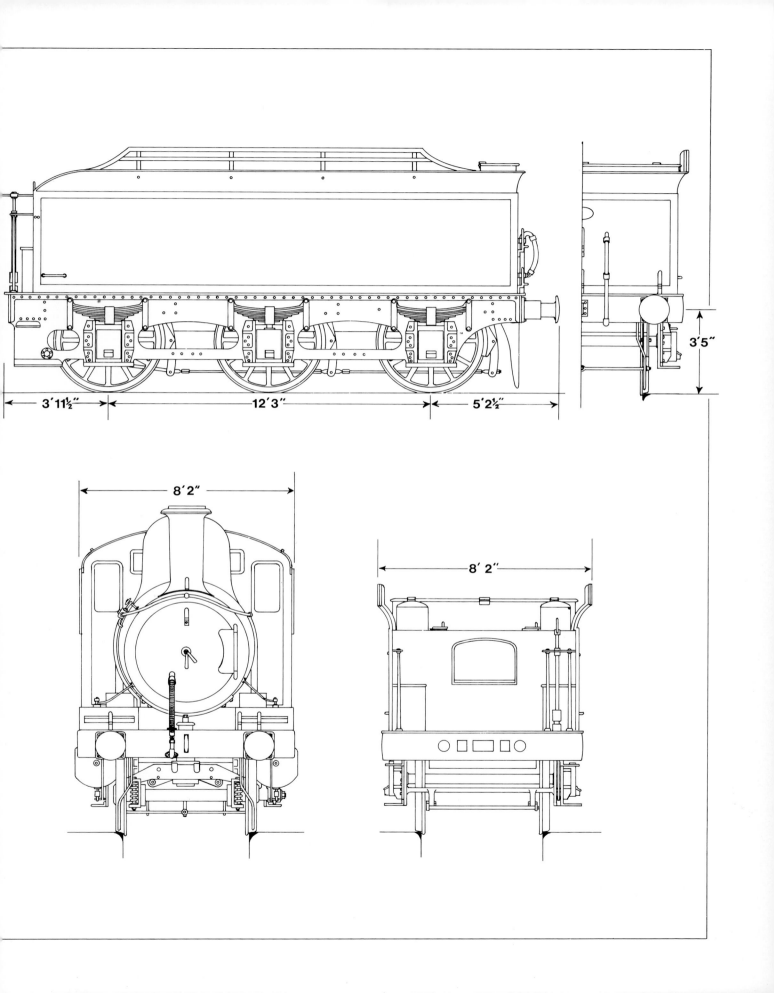

3′5″

3′11½″   12′3″   5′2½″

8′2″

8′2″